Politics of Climate Change
&
Global Peace and Security

Politics of Climate Change & Global Peace and Security

Prof. (Dr.) Vijay Khare
Department of Defence & Strategic Studies,
Dr. Babasaheb Ambedkar Bhavan,
Savitribai Phule Pune University, Pune

PENTAGON PRESS LLP

Politics of Climate Change and Global Peace and Security

Prof. (Dr.) Vijay Khare

First Published in 2021

ISBN 978-93-90095-21-6

Published by
PENTAGON PRESS LLP
206, Peacock Lane, Shahpur Jat
New Delhi-110049
Phones: 011-64706243, 26491568
Telefax: 011-26490600
email: rajan@pentagonpress.in
website: www.pentagonpress.in

Printed at Aegean Offset Printers, Greater Noida, U.P.

Dedicated to
Lord Gautam Buddha
who taught us
Global Peace and Security

CONTENTS

PREFACE

In today's global scenario, the Climate Change has become a contemplative issue that affects the humankind. Through scientific perception, Change in climate patterns means the extreme weather events such as heat waves; flood, storms, drought and bushfires become more frequent, more widespread or more intense. Despite this, repercussions of climate change are far reaching and it will stress our economic, social, and political systems where institutions and government will be unable to manage the stress or endure the shocks of a changing climate. Consequently, the risks to the stability of states and societies will increase. The earth herself faces several global threats in sphere of environment and ecologically she faces imbalance. Climate change is already happening and it represents one of the greatest environmental, social and economic threats facing the planet and raising the question marks on national security

The May 2010 version of the National Security Strategy repeatedly groups together violent extremism, nuclear weapons, climate change, pandemic disease, and economic instability as security threats that require strength at home and international cooperation to address adequately. Climate change is already happening and represents one of the greatest environmental, social and economic threats facing the planet. The United States of America and India are committed to working constructively for a global agreement to control climate change and are leading the way by taking ambitious actions of its own. According to IPCC (2007), climate change is considered to be the major driving force that can trigger alterations in the regional, local and global weather and climate. These changes can in turn, affect the socio-economic, cultural, and political spheres of human societies. The book basically deals with understanding the concept of climate change by examining the related literature.

The introductory chapter gives an overview of entire book which provides readers an understanding about Climate change and its implications on global peace and security. It embarks the quote of former US President Barak Obama, which conveys the importance of Climate change as it affects every aspect of human life across the globe. It also helps readers to understand why it is necessary to tackle climate change issue in current scenario.

Chapter II undertakes a review of literature and methodological approach to the present work. It has incorporated various views and ideas on the Climate Change shared by eminent scholars and experts.

Chapter III discusses the conceptual framework of Climate Change. Under which the global negotiations, summits, conventions and security discourse of Climate Change have been explicated broadly.

Chapter IV namely Politics of Climate Change deals with the impact of global politics on Climate Change and also focuses on national and global action plan. It critically examines various international approaches meant to meet the long term goals of Climate Change.

In chapter V author has tried to portray the relevance of Pandemic and Climate Change by taking COVID-19 issue into consideration which is one of the greatest threats to human and environmental security. The chapter also discusses many contemporary issues like Indian perspectives on national security and pandemic, environment and public health etc.

Chapter VI, core of the book analyses India-US Strategic partnership and Perceptions on Climate Change in a comprehensive manner. It thoroughly discusses India-US Strategic Partnership and Climate Change. It also examines the comparative study of India and US and their diverse perceptions on Climate change.

Chapter VII has widely discussed the role of global actors and international agencies and their responses to avoid serious implications of Climate Change.

The final chapter concludes the overall study of Climate Change and its implications on global peace and security.

Initially, the present work is a result of research project entitled "India-USA Strategic Partnership for tackling Climate Change and its Impact on Global Peace and Security" submitted to the University Grant Commission (UGC), New Delhi. I would like to thank the UGC for its constant support and financial assistance to complete the project. I pay my sincere thanks to my research team and their valuable suggestions and recommendations for completing this book.

I'm thankful to Prof. (Dr.) Nitin R. Karmalkar, Hon'ble Vice Chancellor, Savitribai Phule Pune University for his constant encouragement and kind support for completing this book.

I'm extremely thankful to Prof. (Dr.) S.K. Patil, Former Librarian and Head, Department of Library and Information Science Savitribai Phule Pune University, Dr. Shushma Bhosale, Dr. Deepali Khaire, Dr. Savita Patil, Dr. Anjani Devi, Dr. Ravi Thorat, Mr. P.J. Khache, Dr. Dimple Choudhary, Ms. Bharti Kokil, Ms. Raisa Choudhary, Mr. Satish Gadadare, Mr.Gaurav Kokil, Mr. Onkar Sonawane, Balaji Kamble, Ms. Shivani S. Singh, Sagar Gaikawad, Ms. Sheetal Jagtap, Mr. Dnyaneshwar Gaikawad, Mr. Vijay Mohad, Mr. Ganesh Shinde, Mr. B. Munnuswami and my faculty members of DDSS, and all serving officers who are students and research scholars in the department who helped me a lot in completing this book.

I'm thankful to Mr. Rajan Arya and his team from Pentagon Press LLP for bringing out the book within the shortest possible time. I hope that the readers will find out my efforts worthy and the book will serve the expected needs.

Finally, I express my deep sense of gratitude to my parents, my all time inspiration my wife late Dr. Lalita Khare, my son Dr. Abhishek who is always with me to understand my academic tasks. I record my thanks to everyone who helped directly or indirectly to fulfill my research.

Prof.(Dr.) Vijay S. Khare

1

Introduction

"We are the first generation to feel the impact of climate change and the last generation that can do something about it."

—Former US President, Barack Obama[1]

This research draws on the importance of human security and the climatic changes that affect us all and the respective national and global response to this issue. Highlighting the work of Barry Buzan, with regard to environmental security, he believes that security is an underdeveloped concept and also realised the lack of conceptual literature on security in the 1980s. Being a wide concept, it can be argued that security is a relative term that differs from country to country depending on their priorities and their respective issues. Therefore to develop an in-depth understanding and its various implications in the larger context, the study attempts to examine the existing literature on climate change and its politics by developed states and role of India and the US for tackling climate change. It focuses on the issue of climate change and its impact on global peace and security. Analyses India's role on climate change issue and how the policies framed by the US, impacts on India's national security. Emphasises, how most of the countries are victims of climate change politics. Finally, it enquires how and why the issue of climate change is suddenly looming as a threat to the international security environment. The entire research work is divided into eight chapters and each chapter highlights the major objectives of the study. The chapterisations includes:

1. Introduction
2. Review of Literature and Methodology
3. Conceptual Framework
4. Politics of Climate Change
5. Climate Change and Pandemics
6. India-US Strategic Partnership and Perceptions
7. Politics of Global Actors and Climate Change
8. Conclusion

Broadly, this book focuses on the effect of climate change and how the US and India work to tackle global climate change and the security implications thereof. The US and India have been through a tumultuous relationship in the past, which has improved in the last few decades. Each country has its own methods to tackle climate change on a national level, but globally, the two countries have begun working together to tackle this. Since 9/11, the US and India have also worked closely on security matters. Climate change poses one such existential threat for both the countries. The security of Indians and Americans is of paramount importance to both the countries. The Indian perspective on security largely focused on defence. Over the years, the Indian government has shifted its primary security focus from defence and has also encompassed political, social, economic, technological and strategic aspects. Looking at India's growth trajectory, economically and socially, it can be argued that its consumption of fossil fuels makes it a larger contributing member as a global emitter. Moreover, India, the US, the EU, China and Russia are projected to account for two-thirds of global emissions. One of India's main concerns with regard to climate change has been its poverty. As Indira Gandhi argued, "it is difficult to improve the environment in conditions of poverty. Getting people aware of their individual emissions and its consequential effect on the environment is difficult in a country like India where most of the people were uneducated and living in poverty". Today, with almost all Indians having access to education, it became easier to make them aware of their emissions and encourage restricting those practices.

However, India's arguments have always been on the brink of an 'industrial revolution' and should be allowed the same chance as the Western countries to industrialise without having to worry about the consequential environmental damage. Despite constraints, the Indian government has launched various programs to tackle climate change and industrialize at the same time. The Indian government says that its response to climate change is guided by its own national plans, programs and priorities, and not by the push and pull of international negotiations. Just like the US, India has also encouraged its populous to make a switch to renewable sources from non-renewable ones. The government has started the use of solar, wind and hydro power plants to provide electricity to cities. In smaller villages and towns, the government still uses coal, as it is a cheaper option to provide electricity while industrialising them. The Paris agreement was a turning point for India, where India managed to maintain its diplomatic upper hand over the Western countries by not bowing down to Western pressure to restrict its emissions. The Paris agreement gave India enough room to strive for industrialization with a consciousness towards climate change by scaling its emissions back in terms of vehicles and other sources. The US has been an industrialised country for at least a century. It is widely known and considered to be one of the largest global emitters of greenhouse gases. Through the years, the US failed to ratify most climate change agreements, and each government tried its best to either enhance scrutiny on climate change or reduce scrutiny on climate depending on a Democratic or Republican President respectively.

In the last few decades, the US Environmental Protection Agency (EPA), was entrusted with more power by Congress to check the effects of climate change in the US. While the US was a signatory to almost all climate deals, it never ratified most of the agreements. With the Cancun agreement, the US pledged to reduce its emissions by 17 per cent below its 2005 levels, by helping develop mechanisms to support the needs of developing countries like innovation, research and development, and the innovation of new technologies to track their emissions while they make industrial progress. It can therefore be argued, that the US 'reduced' its emissions by helping other countries increase their emissions to satisfy the needs of the American populous. Under the Obama administration, the Environmental Protection Agency (EPA) flourished by implementing the Clean Air Act and the Clean Water Act. The administration empowered the EPA with special powers to track the emissions from manufacturers and vehicles and check their emissions. The EPA was also instrumental in ensuring that none of their devices are tampered with to check emissions. President Obama himself was aware of the dangers of climate change and pushed for international agreements to be ratified by his administration. He also encouraged manufacturers to use renewable sources of energy instead of fossil fuels. Many NGOs and other supporters of policies for tackling climate change were vocally supportive of measures taken by Obama's administration.

With the election of President Trump, many Green Party supporters and NGOs has been apprehensive that he would overturn the policies implemented by the Obama administration to help tackle climate change. Trump is widely regarded as a climate change sceptic and his appointment of Scott Pruitt is one of the most controversial appointments of the current administration. While not a lot is known about Trump's future decision regarding climate change, it would indeed be a shock for the Americans and the other governments if he decides to deregulate the EPA. There has been a shift in the UN as well, with regard to security. The United Nation (UN) today focuses equally on non-traditional threats to global security, such as hunger, disease, unemployment, crime, social conflicts, political repression and environmental issues. In order to tackle climate change, the UN works through the United Nations Development Programme (UNDP) and the United Nations Environment Programme (UNEP) to bring member states together to tackle climate change. It does so through the Conference of Parties (COP). This book explores the various COPs from Kyoto through Paris. The importance of these agreements will be stressed on. Moreover, the reader will be well informed on each of the agreements on climate change as well as the UN action to tackle climate change. The Paris agreement is widely considered to be a global success as most of the member states have ratified it and have pledged to help restrict the rising temperature to 2°C in this century. It is the responsibility of the COP to ensure that all member states respect and ratify the agreements relating to climate change.

This book will thus, look at all the aspects of security and the threat posed to it by climate change as well as the decisions of the Indian and

American governments to tackle it. Climate is the long-term statistical expression of short-term weather. Climate can be defined as "expected weather". It can be defined by the differences between average weather conditions at two separate times. Climate may change in different time scales. Climate is usually defined as the "average weather" of a place over a long period of time. Climate patterns play a fundamental role in shaping natural ecosystems, and the human economies and cultures that depend on them. But the climate we've come to expect is not what it used to be, because the past is no longer a reliable predictor of the future. Our climate is rapidly changing with disruptive impacts, and that change is progressing faster than any seen in the last 2,000 years.[2]

According to the report, "Preparing for a Changing Climate", rising levels of carbon dioxide and other heat-trapping gases in the atmosphere have warmed the earth and are causing wide-ranging impacts, including rising sea levels; melting snow and ice; more extreme heat events, fires and drought; and more severe storms, rainfall and floods. Scientists project that these trends will continue and in some cases accelerate, posing significant risks to human health, our forests, agriculture, freshwater supplies, coastlines, and other natural resources that are vital to Washington state's economy, environment, and our quality of life.[3] Our climate is powered by the sun and modified by the atmosphere, which is made up of various gases. Some of the gases in the atmosphere allow sunlight to pass through, and prevent the heat from escaping back out into space, similar to the glass in a greenhouse. This is called greenhouse effect. The gases such as water vapour, carbon dioxide and methane are called greenhouse gases. The earth's climate is a result of complex processes and interactions between the atmospheres (including the greenhouse effect), oceans, land, ice, plants and animal (including humans).

- Life on earth is possible because of the warmth of the sun. While some of this incoming solar radiation bounces back into space, a small portion of it is trapped by the delicate balance of gases that make up our atmosphere. Without this layer of insulation, earth would simply be another frozen rock hurting through space. Carbon dioxide (CO_2) is the most important gas in this layer of insulation.

- Carbon is stored all over the planet in plants, soil, the ocean and even us. We release it into the atmosphere as carbon dioxide through activities, such as burning fossil fuels (coal, oil and gas) and cutting down trees. As a result, today's atmosphere contains 42 percent more carbon dioxide than it did before the industrial era.
- We have released so much carbon dioxide and other greenhouse gases that our planet's atmosphere is now like a thick, heat-trapping blanket. By disrupting the atmospheric balance that keeps the climate stable, we are now seeing extreme effects around the globe. It is like a thermostat that's gone haywire and it just doesn't work the way it should. The result: the climate changes and it gets warmer. Extreme weather events also become more common.
- Global warming has already begun. Since 1900, the global average temperature has risen by 0.7°C and the northern hemisphere is substantially warmer than at any point during the past 1,000 years.

Our understanding of climate change is largely the result of the Intergovernmental Panel on Climate Change (IPCC), the world's most authoritative voice on the topic. Because so many systems are tied to climate, a change in climate can affect many related aspects of where and how people, plants and animals live, such as food production, availability and use of water and health risks. For examples, a change in the usual timing of rains or temperatures can affect when plants bloom and set fruit, when insects hatch or when streams are at their fullest. This can affect historically synchronized pollination of crops, food for migrating birds, spawning of fish, water supplies for drinking and irrigation, forest health, and more. Drastic changes in global climate patterns will pose a direct threat to peace and security around the world, warns the United Nations Secretary-General. Extreme weather events continue to grow more frequent and intense in rich and poor countries alike, not only devastating lives, but also infrastructure. In the last 100 years or so, the Earth's surface and lowest part of the atmosphere have warmed up an average by about 0.6°C during this period, largely as a result of the burning of fossil fuels for energy and transportation and land use changes, for food by man-kind. In the last 20 years, concern has grown that these two phenomena are, at least in part, associated most probably to the man-made increases in greenhouse gas emission. While some factors that contribute to our current climate change are manmade there are some natural causes of climate change too, including changes in the amount of energy coming from the sun, volcanic venting and shifting patterns of ocean circulation, can cause global climate to change over similar periods of time. The balance of evidence indicates that there is a discernible human influence on the global climate. "Climate change has long since ceased to be a scientific curiosity, and is no longer just one of many environmental and regulatory concerns. As the United Secretary General has said, it is the major, overriding environmental issue of our time, and the single greatest challenge facing environmental regulators. It is growing crisis with economic, health and safety, food production, security and other dimensions".

Viewing the Earth's greenhouse effect and its consequences as a causal chain six major links can be highlighted:

- The social driving forces of human activities that produce greenhouse gas emissions.
- Greenhouse gas emissions that provide the enhanced radioactive.
- Forcing that induces climate change, which impacts nature and society.
- Finally, the anticipation and experience of the effects of climate change encourage a range of human responses to prevent climate change, mitigate it, or adapt to it.

Climate change occurs overtime, and the degree of change depends on the time scale we choose to examine. Changes in the last decade seem insignificant when compared to those over the past million years. Long-term natural changes resulted from volcanic activity releasing huge quantities of heat-trapping CO_2 into the atmosphere and from the evaluation of plants that remove CO_2 from the atmosphere through photosynthesis. Climate also undergoes natural cooling and warming cycles as a result of periodic fluctuations in the amount of solar energy reaching the Earth's surface. These fluctuations occur on a scale of tens of thousands of years and result from periodic changes in the orbital alignment of the Earth with respect to the sun, known as Milankovitch Cycles. Paleo climatologists use many methods to determine past climates.

These include (from long past to recent) fossil rocks, isotope ratios in plankton and coral reefs, polar ice cores, plant pollen, tree rings, historical documents and scientific instrumental records. About 11,000 years ago, melting glaciers lowered the density of seawater in the North Atlantic. This slowed oceanic conveyor belt carrying heat to Europe and North America. This event called, "The Younger Dryas," demonstrated that climate can change rapidly – in response to an environmental disturbance. Thus, natural fluctuation in the global average temperature of the Earth (2 to 3°C) occurs on a timescale of tens of thousands of years. These long-term natural fluctuations have been accompanied by major changes in ecosystems and in recorded history have influenced human civilizations in a number of ways. Now, human emissions of greenhouse gases threaten to raise the Earth's global average temperature by 2°C to 6°C or even more during this century. Climate change is a significant and long lasting change in the weather patterns over periods ranging from decades to several years. This can be a change in average weather conditions, or in the distribution of weather around the average conditions.

According to United States Environmental Protection Agency (USEPA) the earth's average temperature has risen by 0.8°C (1.4°F) over the past century, and is projected to rise another 1.1 and 6.4°C (2 to 11.5°F) over the next hundred years. These small changes in the average temperature of the planet can translate to large and potentially dangerous shifts in climate and weather. According to EPZ (2010) climate change is a term that refers to major changes in temperature, rainfall, snow, or wind patterns lasting for decades or longer. Both human-made and natural factors contribute to climate change:

- Human causes include burning fossil fuels, cutting down forests, and developing land for farms, cities, and roads. These activities all release greenhouse gases into the atmosphere.
- Natural causes include changes in the Earth's orbit, the sun's

intensity, the circulation of the ocean and the atmosphere, and volcanic activity.

The warming of the climate system is unequivocal as is now evident from observations of increase in global average air and ocean temperatures, widespread melting of snow and ice, and rising global mean sea level. The earth's average surface temperature has risen by 0.76°C since 1850. Most of the warming that has occurred over the last 50 years is very likely to have been caused by human activities. In its Fourth Assessment Report (AR4) published in 2007, the Intergovernmental Panel on Climate Change (IPCC)[4] projects that, without further action to reduce greenhouse gas emissions, the global average surface temperatures is likely to rise by a further 1.8-4.0°C this century and by up to 6.4°C in the worst case scenario. Even the lower end of this range would take the temperature increase since pre-industrial times above 2°C the threshold beyond which irreversible and possibly catastrophic changes become far more likely. Projected global warming this century is likely to trigger serious consequences for mankind and other life forms, including a rise in sea levels of between 18 to 59 cm which will endanger coastal areas and small islands, and a greater frequency and severity of extreme weather events. The science of climate change has experienced an extraordinary expansion in the period since mid-1980s. This is evidenced particularly by the growth in the number of scientific publications in this field. In the Science Citation Index (SCI) the publication data increases in from 1986-88 when the number of published papers on 'climate change' quadrupled. Part of this interest was sparked by climate scenarios produced from new generations of climate models devised in the late 1980s. Environmental campaigning organizations such as Greenpeace, Friends of the Earth and other opposing groups, such as the Global Climate Coalition in the US and the Road Haulage Association in the UK each quote selectively from scientific studies so as to promote their particular perspectives on the extent of human influence in climate and environmental change.

Climate change and its consequences are being increasingly viewed as the foremost problem of the 21st century. Climate change challenges at the national, regional and global level are hugely demanding and interconnected and have obvious implications in terms of human security. Significantly, and depending on the nature of the changes and other relevant factors, climate change could also lead to intersect and/ or intrastate conflicts. In order to assess the role of climate related factors, particularly in the context of our region, and to provide greater clarity about the potential sources of instability/conflict in the future. Climate change has multidisciplinary dimensions and as noted by the Intergovernmental Panel on Climate Change in its report of 2007 which stated that it could accentuate the conditions of resource scarcity and environmental degradation. In turn these could create fissures in the international system. The resultant challenges are shortages of water, food grains and loss of biodiversity, collapse of ecosystems and frequent disasters and migration. The north-south divide has marked the climate change debate and negotiations. The industrialized developed countries tend to securitize climate change so as to apply pressure on developing countries to cap greenhouse gas emissions. The developing countries on the other hand, want the issue to be addressed principally from an economic and social perspective under the UN General Assembly and not the Security Council.

Uttarakhand received heavy rainfall in the month of June 2013 (14 to 17 June 2013), which was about 375 percent more than the benchmark rainfall during a normal monsoon in India, which lead to the emergence of heavy and serious floods in Uttarakhand, Himachal Pradesh and Western Nepal. It also caused acute rainfall in other nearby regions of Delhi, Haryana, and Uttar Pradesh and some parts of Tibet. Heavy rainfall for four consecutive days as well as melting snow aggravated the floods and resulted in serious damage of life and property.

The scientific discovery of climate change had begun in the early 19th century when ice ages and other natural changes in pale climate were first identified and the phenomenon of natural greenhouse effect was identified.

During the 1960s, although some scientists had pointed out that human activities in the form of atmospheric aerosols (for e.g. pollution) could even have cooling effects but still the warming effect due to the presence of carbon dioxide gas had become increasingly significant and convincing. The scientific opinion increasingly favoured the warming viewpoint in the 1970s and by the 1990s it was noticed that the greenhouse gases were deeply involved in most climate changes and emissions due to human activities causing serious global warming. Due to the improving fidelity of computer models and observational work confirming the Milankovitch theory of the ice ages a consensus position was formed.

NOTES

1. The White House, Office of the Press Secretary, Remarks by the President at U.N. Climate Change accessed through Summit https://obamawhitehouse.archives.gov/the-press-office/2014/09/23/remarks-president-un-climate-change-summit on 5 August 2019.
2. Department of Ecology State of Washington, What is Climate Change?, Accessed through http://www.ecy.wa.gov/climatechange/whatis.htm on 1 November2016.
3. Ibid.
4. IPCC, The intergovernmental panel on climate change, Working Group III Climate Change 2007: Mitigation of climate change May 2007. Accessed through http://www.ipcc.ch/pdf/assessment-report/ar4/wg3/ar4_wg3_full_report.pdf on 17 November 2017.

2

Methodology

2.1 Review of Literature

The environment is changing rapidly. Increase in world population, accompanied by industrialization and other human activities, altering the atmosphere, ocean, land, ice cover, ecosystems, and the distribution of species over the planet. Understanding these and other global changes, including climate change, is critical to our nation's health and economic vitality. Scientific research is critical to gaining this understanding. Research, along with an array of increasingly sophisticated tools for collecting and analyzing data, can provide essential knowledge to governments, businesses, and communities as they plan for and respond to the myriad manifestations of global change, including sea-level rise and ocean acidification, heatwaves and drought, and the severe storms, floods, and forest fires that pose an ever-growing risk to life, property and agriculture. Making the best decisions with regard to global change will depend on a well-grounded understanding of the Earth system and the changes taking place within it. Therefore, the study explores the existing literature to find out the gaps and as it would be vital to identify the emerging trends and shifting nature of climate change to tackle the issue successfully and also come up with the effective policy measures. According to Hegerl, G.C. and Zwiers, F.W.[1] the term 'Climate change' refers to a change in the state of the climate that can be identified (e.g., using statistical tests) by changes in the mean and/or the variability of its properties, and that persists for an extended period, typically decades or longer. Climate change may be due to internal processes and/or external forces. Changes in solar radiation and volcanic activity occur naturally and they contribute to the total natural variability of the climate system, while there are changes in the composition of the atmosphere through emission of gases that are the result of various anthropogenic activities. The major anthropogenic activities contributing to climate change are burning of fossil fuels, deforestation, urbanisation and rapid increase in air emissions due to industrial activities. Because of the complex and dynamic behaviour of nature, climate change can have different impacts in different places/ecosystems and the likelihoods of these impacts can also vary greatly. Some ecosystems might experience only minor changes whereas others might cease to exist, supplanted by new ecosystems. According to the Intergovernmental Panel on Climate Change (IPCC), Carbon dioxide (use of fossil fuels) contributes maximum to global greenhouse gas emissions by gases followed by carbon dioxide (deforestation, biomass decay), methane, nitrous oxide and a small fraction contributed by other gases. The energy supply sector contributes maximum in the category of global greenhouse gas emissions by source followed by industrial sector, forestry, agriculture, transport, residential and commercial buildings and waste and wastewater. The percentage contribution in both these categories is given in Figure 1 and Table 1 summarizes the international responses to climate change.

According to EPA, Earth's average temperature has risen by 0.8°C (1.4°F) over the past century, and is projected to rise another 1.1 and 6.4°C (2 to 11.5°F) over the next hundred years. Large and potentially dangerous shifts in climate and weather can occur because of even small changes in the average temperature of the planet. Rising global temperatures have been accompanied by changes in weather and climate and changes in rainfall, resulting in more floods, droughts, or intense rain as well as more frequent and severe heat waves being noticed in many areas. The planet's oceans and glaciers have also experienced some big changes oceans are warming and becoming more acidic, ice caps are melting, and sea levels are rising. As these and other changes become more pronounced in the coming decades, they will likely present challenges to our society and our environment.

Figure 1 (A): Global Greenhouse Gas Emissions by Gases

Figure 1 (B): Global Greenhouse Gas Emissions by Source

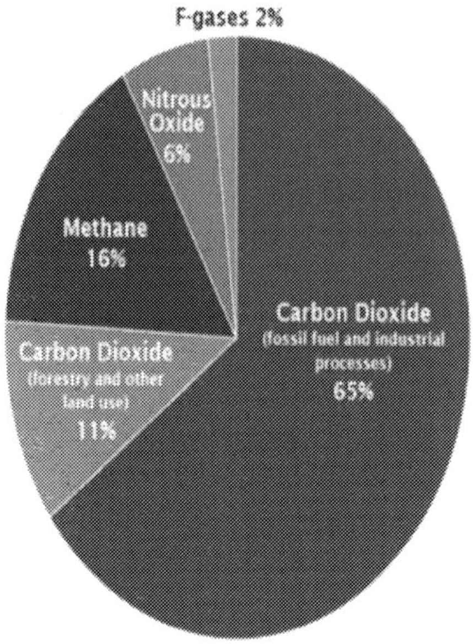

A

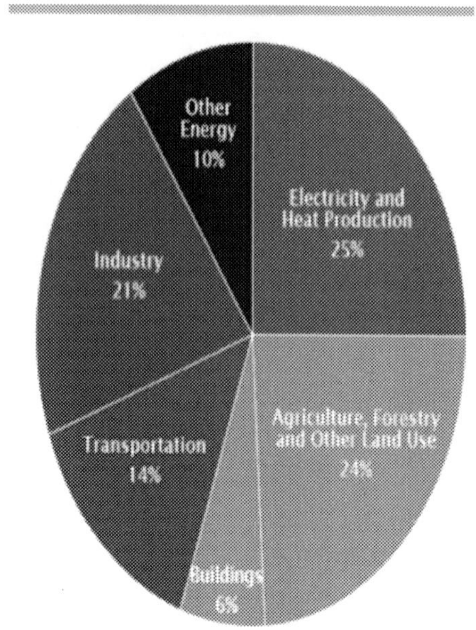

B

Source: IPCC, 2015.

According to IPCC (2015) between 1970 and 2004 the world's greenhouse gas emissions rose by 70 percent and in recent years many countries have introduced policies to reduce them. Despite these so-called 'mitigation policies', emissions are predicted to be 45-110 percentage higher in 2030 than they were in 2000 a result of projected trends in economic growth, consumer and business behaviour, and the number of people living on the planet. Rise in sea level, melting of ice glaciers, damage to ecosystems, spread of diseases, rise in surface temperature, biodiversity loss, etc., are some of the serious impacts that have resulted because of climate change. These multiple impacts of climate change will result in damage to the environmental sustainability and would ultimately question the survival of human life on this beautiful planet. The overall impacts that can be caused due to climate change are summarized in Figure 1 and the entire climate change processes, characteristics and threats as given by UNFCCC is given in Figure 3 as follows:

2.1.1 Impacts of Climate Change

According to McKibbin W.J. the global community has been struggling with the issue of how to effectively respond to the threat of climate change for several decades. In 1992, the United Nations Earth Summit in Rio de Janeiro produced a landmark treaty on climate change that undertook to stabilize greenhouse gas concentrations in the atmosphere. By focusing on stabilization, however, the treaty implicitly adopted the position that the risks posed by climate change require that emissions be reduced no matter what the cost. The agreement, signed and ratified by more than 186 countries, including the United States, was a result of numerous subsequent rounds of climate negotiations and aimed at rolling back emissions from industrialized countries to the levels that prevailed in 1990. The classic example of the impacts of climate change can be seen

Flow Chart 1: Global Impacts of Climate Change

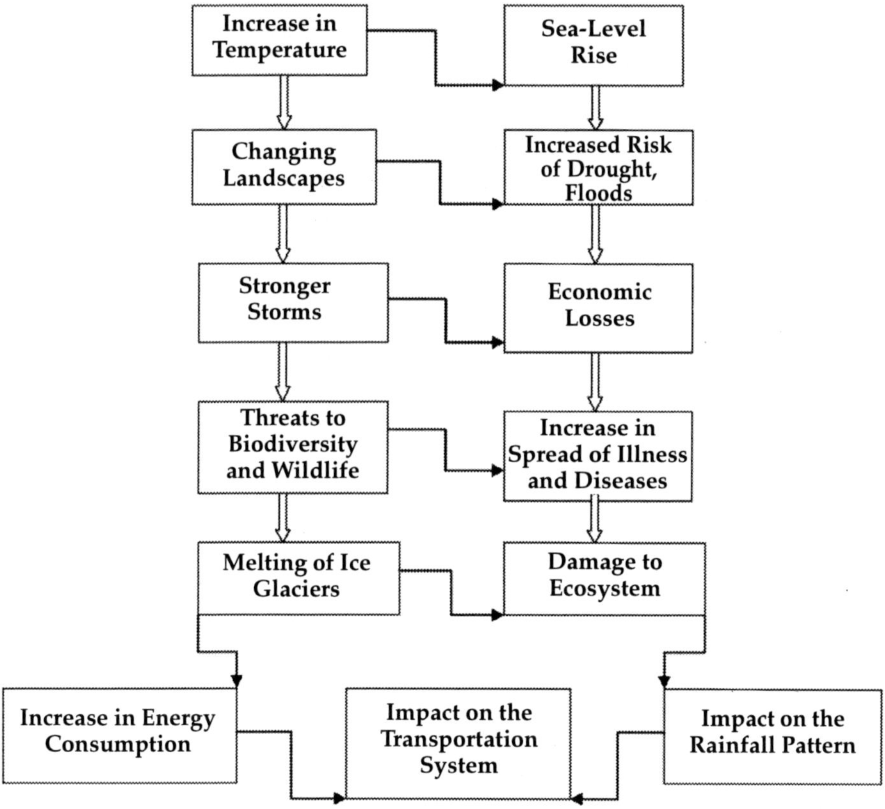

from the disappearance of ice from Mount Kilimanjaro as shown in Figure 3.

Figure 2: Climate Change: Process, Characteristics and Threats

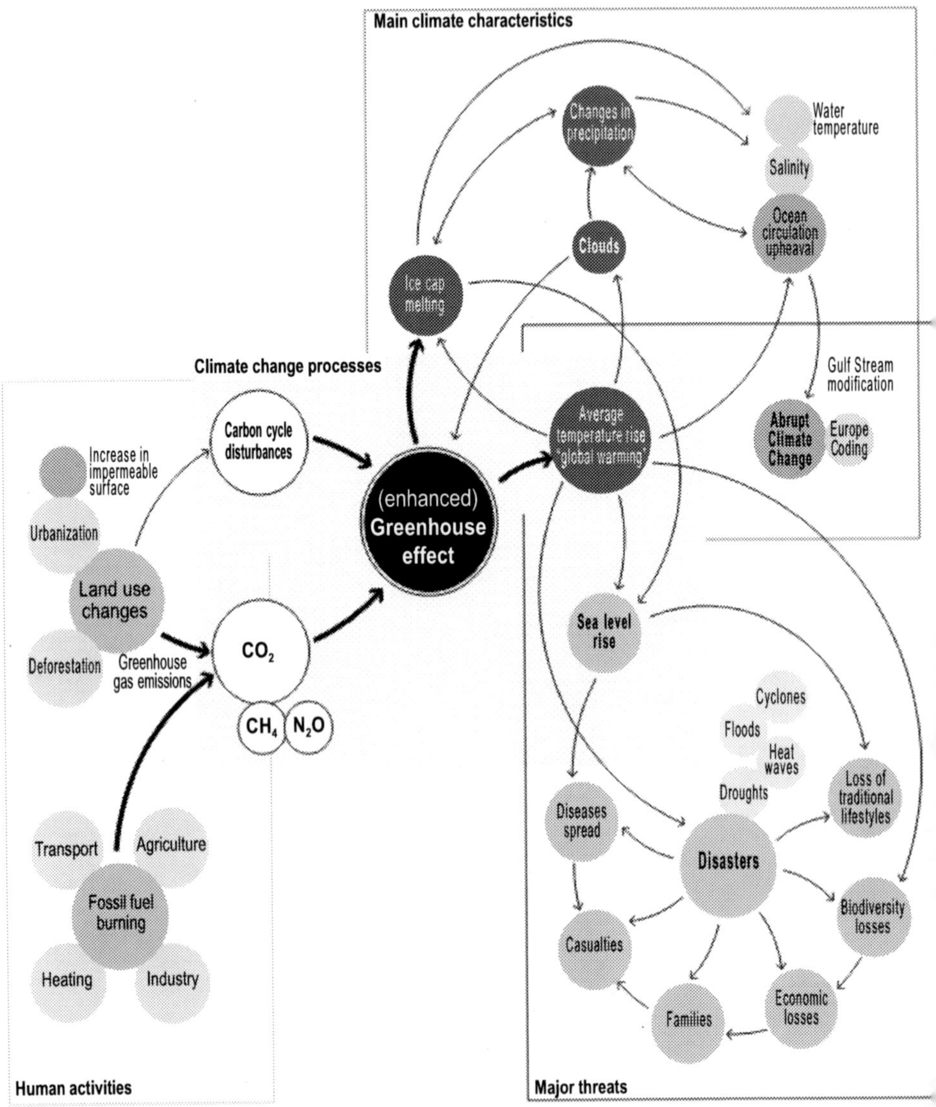

Source: UNFCCC, 2019.

Figure 3: The retreat of glaciers on Mount Kilimanjaro can be seen in the photographs from 1912, 1970, 2000, and 2006; from 1912 to 2006, 85 per cent of the ice has disappeared

Source: Thompson L, 2010.

Asia is the largest continent on the Earth and spreads over four climatic zones (boreal, arid and semi-arid, tropical and temperate) and it faces formidable environmental and socio-economic challenges in its effort to protect valuable natural resources. Land and ecosystems are being degraded, threatening to undermine food security and the water and air quality are deteriorating while continued increases in consumption and associated waste have contributed to the exponential growth in the region's existing environmental problems[2]. The regional impacts and vulnerabilities to climate change in Asia have been summarized in Table 1.

Figure 4: The link between greenhouse gases and climate change Carbon dioxide Emissions from Selected Countries

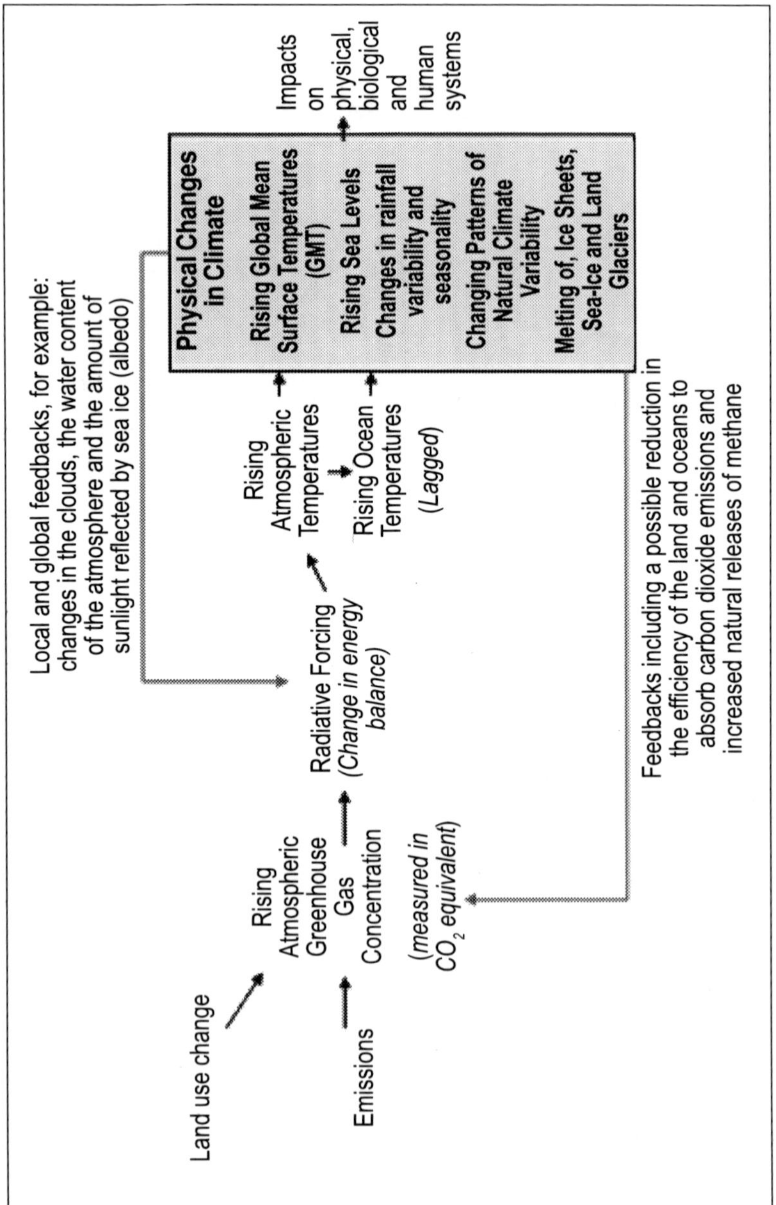

Table 1: Regional Impacts and Vulnerabilities to Climate Change in Asia (UNFCCC 2007)

Impacts	*Sectoral Vulnerabilities*	*Adaptive Capacity*
Temperature • Warming above the global mean in central Asia, the Tibetan Plateau, northern, eastern and southern Asia. Warming similar to the global mean in Southeast Asia. • Fewer very cold days in East Asia and South Asia ***Precipitation, snow and ice*** • Increase in precipitation in most of Asia. Decrease in precipitation in central Asia in summer. • Increase in the frequency of intense precipitation events in parts of South Asia, and in East Asia. • Increasing reduction in snow and ice in Himalayan and Tibetan Plateau glaciers ***Extreme Events*** Increasing frequency and intensity of extreme events particularly: • droughts during the summer months and El Niño events; • increase in extreme rainfall and winds associated with tropical cyclones in East Asia, Southeast Asia and South Asia; • intense rainfall events causing landslides and severe floods; • heat waves/hot spells in summer of longer duration, more intense and more frequent, particularly in East Asia.	***Water*** • Increasing water stress to over a hundred million people due to decrease of freshwater availability in Central, South, East and Southeast Asia, particularly in large river basins such as Changjiang. • Increase in the number and severity of glacial melt-related floods, slope destabilization followed by decrease in river flows as glaciers disappear. ***Agriculture and food security*** • Decreases in crop yield for many parts of Asia putting many millions of people at risk from hunger. • Reduced soil moisture and evapotranspiration may increase land degradation and desertification. • Agriculture may expand in productivity in northern areas. ***Health*** • Heat stress and changing patterns in the occurrence of disease vectors affecting health. • Increases in endemic morbidity and mortality due to diarrheal disease in south and Southeast Asia. • Increase in the abundance and/or toxicity of cholera in south Asia. ***Terrestrial Ecosystems*** • Increased risk of extinction for many species due to the synergistic effects of climate change and habitat fragmentation. • Northward shift in the extent of boreal forest in north Asia, although likely increase in frequency and extent of forest fires could limit forest expansion. ***Coastal Zones*** • Tens of millions of people in low-lying coastal areas of south and Southeast Asia affected by sea level rise and an increase in the intensity of tropical cyclones. • Coastal inundation is likely to seriously affect the aquaculture industry and infrastructure particularly in heavily-populated megadeltas. Stability of wetlands, mangroves, and coral reefs increasingly threatened.	Adaptive capacity varies between countries depending on social structure, culture, economic capacity, geography and level of environmental degradation. Capacity is increasing in some parts of Asia, for example the success of early warning systems for extreme weather events in Bangladesh and the Philippines. However, capacity is still constrained due to poor resource bases, inequalities in income, weak institutions and limited technology.

According to Stern Review, the atmosphere is much more complicated than what these simple models suggest. The resulting warming will in fact be much greater than 1°C because of the interaction between feedback in the atmosphere that acts to amplify or dampen the direct warming.

There has been significant increase in the global emission of greenhouse gases from 1990 to till today. The emissions from developed countries like US and UK have always been above 3500 MT CO_2 from 1990 to 2006. The greenhouse gas emissions from China were around 1800 MT CO_2 in 1990 and it reached above 5250 MT CO_2 in 2006. The greenhouse gas emissions from India are much less as compared to US, UK and China.

Figure 5: Growth of Total CO_2 Emissions of Selected Countries

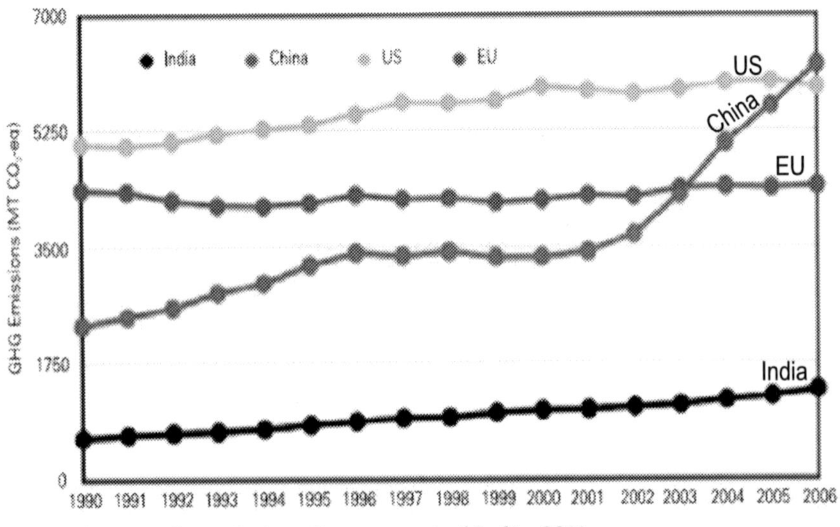

Source: Planning Commission, Government of India, 2011.

According to EPA (1983), emission of carbon dioxide, methane and other gases intensifies the greenhouse effect of the atmosphere and thus causes long-term changes in the climate (like rise in temperature). Some of the gases in the earth's atmosphere more particularly water vapour and carbon dioxide have an ability to absorb infrared radiation and they do not prevent sunlight from reaching the surface of the earth and warming it, but they do trap some of the infrared radiation emitted back into the space by the earth, i.e. they work in a way similar to the glass in a greenhouse. Greenhouse gases have always been present in the atmosphere but now their concentration is rising as a result of anthropogenic activities, which are indirectly increasing the greenhouse effect.

Earth's atmosphere is composed of many different gases, three of which make up about 99-95 per cent of the atmosphere nitrogen (78.08 per cent) oxygen (20.95 per cent) and argon (0.92 per cent). However, it is the other gases that make up 0.05 per cent of the atmosphere (trace gases) that are of most concern. The two trace gases, namely carbon dioxide and methane, play a vital role in the atmosphere. While being virtually transparent to the shorter wavelengths of energy received from the sun, these gases absorb the longer wavelengths of energy given off by the earth. Thus a portion of the energy from the earth that would normally be released to space is absorbed by these trace gases and is then transmitted back to the earth. Hence, these gases are known as greenhouse gases i.e., they allow the sunlight but they prevent much of the energy from radiating back through.

Figure 6: The Link between Greenhouse Gases and Temperature, 1850–2009

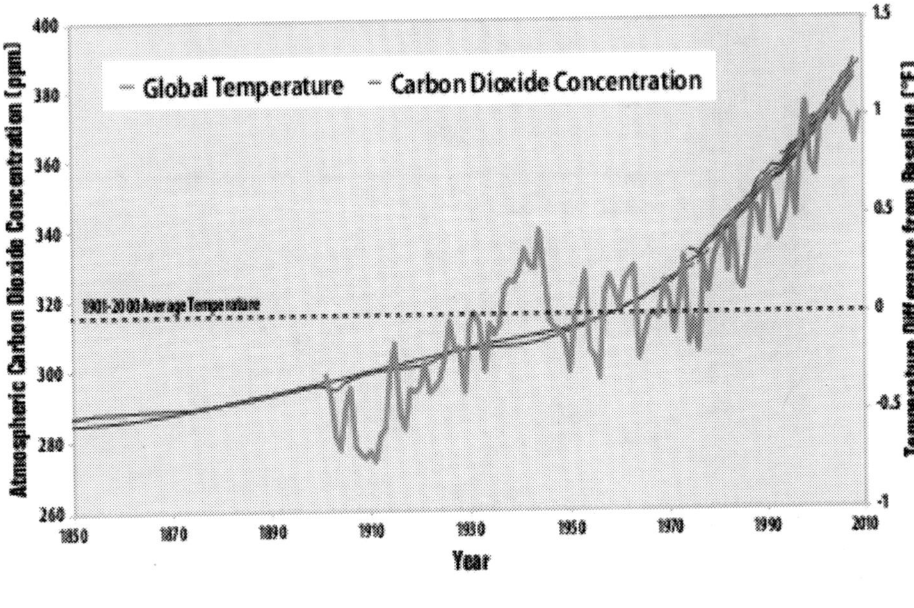

According to EPA,[3] the emissions of carbon dioxide, which is an important greenhouse gas, has been increasing since the industrial revolution and it is causing carbon dioxide levels to build up in the atmosphere and global temperatures to rise. The temperatures have gone up at an increased rate in particular over the past 30 years. The link between greenhouse gases and temperature from 1850–2009 is shown in figure 6. The temperature data shows a great difference from the average or baseline temperatures between the years 1901 and 2000.

Based on the data from IPCC,[4] the globally observed average combined land and ocean surface temperatures anomaly from 1850-2012, and the observed change in surface temperature from 1901-2012 are shown in Figure 7 (A) and (B) respectively.

Figure 7(A): Observed globally averaged combined land and ocean surface temperature anomaly 1850-2012

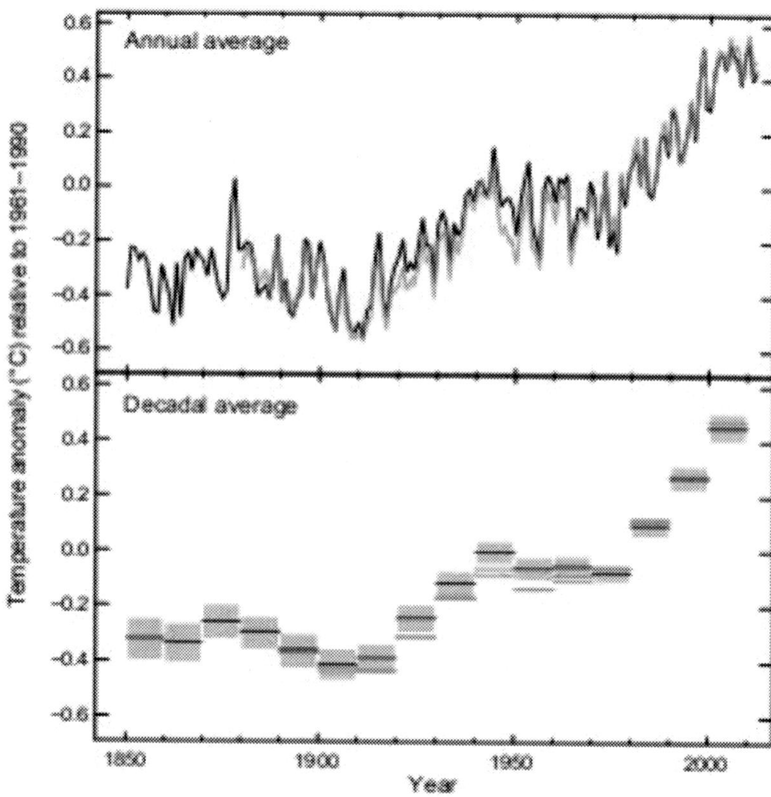

Figure 7(B): Observed change in surface temperature 1901-2012

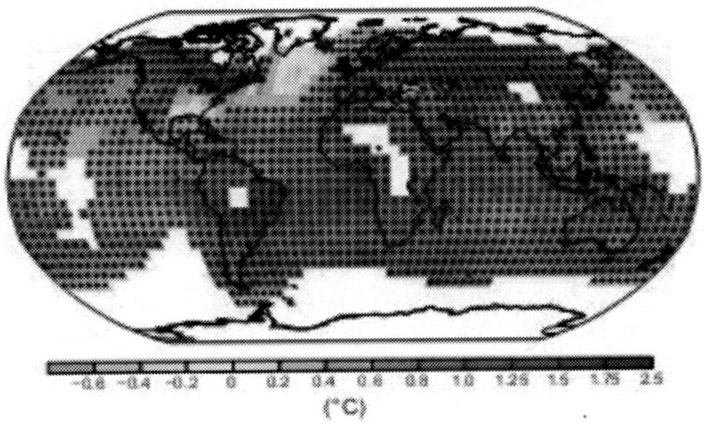

Source: IPCC 2013.

Figure 7(A) observed global mean combined land and ocean surface temperature anomalies, from 1850 to 2012 from three data sets. Top panel: annual mean values. Bottom panel: decadal mean values including the estimate of uncertainty for one dataset (black). Anomalies are relative to the mean of 1961-1990.

Figure 7(B) Map of the observed surface temperature change from 1901 to 2012 derived from temperature trends determined by linear regression from one dataset (orange line in panel a). Trends have been calculated where data availability permits a robust estimate (i.e. only for grid boxes with greater than 70 per cent complete records and more than 20 per cent data availability in the first and last 10 per cent of the time period). Other areas are white. Grid boxes where the trend is significant at the 10 per cent level are indicated by a + sign.

Figure 8 show variety of temperature records over the last 1000 years based on a variety of proxy recorders, such as tree rings, ice cores, historical records, instrumental data, etc., show the extent of the recent warming. The shaded region shows the range of temperature projected by Meehl et al. (2007) to 2100 AD, and the filled circle depicts the average of the range.[5]

Figure 8: Variety of temperature records over the last 1000 years based on a variety of proxy recorders

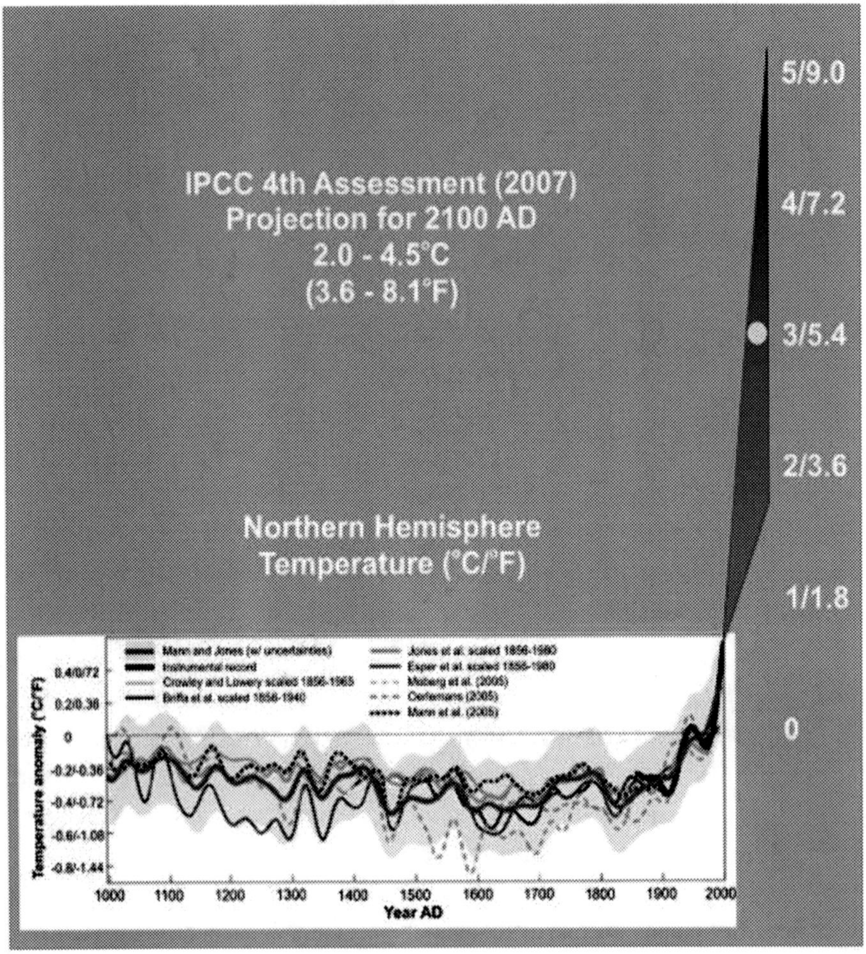

Tubi A, Fischhendler I and Feitelson E[6] assessed the degree to which vulnerability may improve the level of explanation of mitigation policies that were adopted by studying over 90 countries between 1990 and 2011. They found that mitigation and adaptation are not viewed as substitutes by high adaptive capacity countries and the insignificance of the impacts is generally caused by the uncertainty in their assessment.

Barnett J explored the range of possible connections between climate change and security[7] and he showed the possible ways in which climate change is a security issue. Parry M[8] investigated the key areas on impacts, vulnerabilities and adaptation to climate change and concluded that there is a special need to increase the observational and research capacities particularly in the developing countries. Klein et al,[9] showed that the objective of climate policy should be to facilitate development and implementation as part of sectorial policies and showed focus on the optimal use and expected effectiveness of financial instruments.

According to Laukkonen J et al,[10] climate change is a global phenomenon that impacts the different societies throughout different scales, ranging from individuals to localities and even the entire region. They concluded that a methodology and a tool needs to be developed that can help individuals, regions or countries in the decision-making process towards climate change.

Martens et al.,[11] suggested that the science of integrated assessment combined with its application of scenario, modelling and participatory approaches can have considerable potential for analysing the multiple causes and impacts of a complex problem like climate change and to inform the development of effective policy responses.

According to Lorenzoni et al.,[12] climate and social scenarios can be integrated for understanding the inter-relationships between a changing climate and the dynamic evolution of social, economic and political systems. They also showed that these integrated scenarios could be calibrated so that they can be applied 'bottom up' to local stakeholders in vulnerable sectors of the economy.

According to Tol et al.,[13] poorer people of the world are likely to be affected by the impacts of climate change because they are exposed more to the weather, they are closer to the biophysical and experience the limits of climate and also their adaptive capacity is lower. They also concluded that the impacts of climate change are substantial but subtle, and the mitigation and adaptation policies for the complex problem can be guided properly through more research.

Ikeme J[14] provided a mapping of the different takes on the moral aspects of environmental decisions particularly on the twin concepts of equity and environmental justice. A more concrete way on how the two concepts should be constructed in evaluating propositions towards tackling global environmental issues and understanding the source of North-South debate on distribution of climate change burdens was also demonstrated.

Kaechele et al.,[15] elaborated the approach and flexibility of carbon trading, analyzed the emission profile of rural India and the significance of its future emission pathways within the proposed framework. They also showed how low carbon development in India can assist in cost-effective decarbonisation of industrialised countries and mitigation of climate change.

Rong F[16] undertook comparative analysis of Brazil, China, India, Mexico and South Africa for understanding the stances of developing countries on post-2012 climate change negotiations and found that mitigation capability is the crucial factor that consists of at least such sub-factors as per capita income, energy endowment and economic structure while ecological vulnerability does not seem to play an important role that includes reductions in agricultural outputs, sea level rise, climate-related natural disasters, etc.

According to Sathaye et al.,[17] the projected climate change under various scenarios is likely to have implications on biodiversity, livelihoods, water supply, food production etc. Indian has a significant stake in scientific advancement as well as an international understanding to promote mitigation and adaptation.

Baettig et al.,[18] presented a method to estimate and compare the cooperative behaviour of countries within the international climate change regime. They gave two indicators to measure whether and how effectively. Over the last 100 years, the average air temperature near the Earth's surface has risen by a little less than 1°C or 1.3 degree Fahrenheit. It may not be seen to a significant rise. Yet this warming is responsible for the conspicuous increase in storms, floods and raging forest fires we have seen in recent years says scientists. Their data shows that an increase of 1°C makes the earth warmer now than it has been for at least a thousand years. The world's five warmest years have all occurred since 2015 with nine of the 10 warmest years occurring since 2005, according to scientists from NOAA's National Centres for Environmental Information (NCEI). It was also the 43rd consecutive year with global land and ocean temperatures, at least nominally, above average. The average temperature across the globe in 2019 was 1.71 degree F (0.95 of a degree Celsius) above the 20th-century average and just 0.07 of a degree F (0.04 of a degree Celsius) cooler than the 2016 record. (Figure 9)

Projections by the UN climate change body, the Intergovernmental Panel on Climate Change say that global surface temperature will probably rise further 1.1 to 6.4°C (2.0 to 11.5° Fahrenheit) during the 21st century. A certain degree of warming is unavoidable even if we manage to reduce our emissions immediately. Oceans for example act as huge heat repositories that follow changes in air temperature with a time lag of decades or even hundreds of years. Melting ice caps reflect less sunlight, so our planet absorbs more and more heat. Exactly how these changes will influence the warming trend is unclear. All we know for certain is that it's going to get warmer and that human greenhouse gas emissions are an important reason for this. Rising sea levels, increasing weather extremes, change in rainfall patterns, disruption of safe drinking water sources and water scarcities in non-monsoonal months pose serious risks to social and political harmony in Southern Asia. Such trends are also likely to influence the vector of disease control and potentially create major public health challenges in this region and beyond. Also, it will be the poor who will be hit the hardest. Furthermore, the impact of climate change will extend beyond human civilization to Southern Asia's exceptionally rich plant and animal world. Today's endangered species could become extinct tomorrow. Meeting the challenges posed by climate change, therefore, demands that sustained efforts begin now. That in turn, means switching to a more climate-friendly path in development and energy needs. The only sure path to energy security, in any event, lies through renewable sources of energy. Renewable also offer clean energy. For the foreseeable future, however, coal will continue to play a major role in meeting the electricity needs in Southern Asia and China. However, India, China and other states need to embrace cleaner technologies like coal gasification that holds immense promise to cut down emissions of carbon dioxide.

Figure 9: A Map of the World Noting some of the most Significant Climate Events of 2019

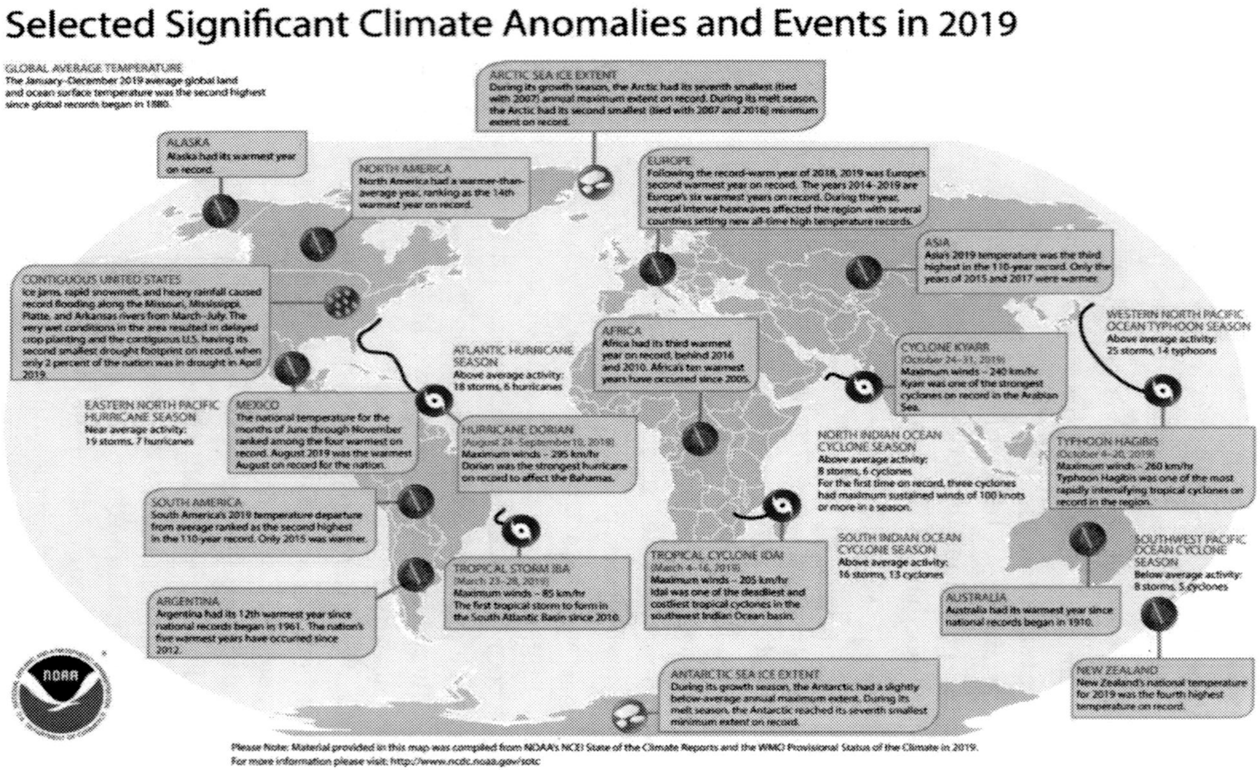

Source: National Centres for Environmental Information, Accessing the Global climate in 2019, Accessed through https://www.ncei.noaa.gov/news/global-climate-201912,on 20 January 2020.

2.2 METHODOLOGY

Climate change like many global problems currently recognized as a threat to the international security and cooperation. In theoretical terms, it is being securitized and included in the traditional security studies. Climate change and its accompanying environmental degradation are perceived to be a threat that can have incalculable consequences on the international community. The United Nations (UN), European Union (EU), United States of America (USA), African Union and many other regions of the world have recently begun to ask how the security dimensions of climate change will affect access to shared resources, national security, borders integrity, trade routes and international political alliances and economic relations. The impacts of climate change will fuel the politics of resentment between those most responsible for climate change and those most affected by it and it will, by acting as a threat multiplier, deepen political tensions nationally and internationally. Therefore, the primary aim of the research is to carry out a detailed study of climate change and strategic cooperation between India and the USA and its impact on international peace and security. The literature on climate change shows that there is no common consensus on climate change issues. The developed countries and developing countries are taking their stands often suiting their own strategic or short-term national interests govern by power politics. The evolving debates on climate change suggest the potential to inquire about the divergence in perspectives which shaping the politics of climate change. The study has been conducted by using the historical method, content analytical method and empirical data which are extensively compiled and reviewed to critically examine the politics of climate change. It is largely based on the secondary sources of data which include national and international reports on climate change.

2.2.1 Limitations of the Study

The issue of climate change needs a scientific and technical approach to deal because conventional method is not adequate to address the issue. It requires developing more interdisciplinary approach to respond the climate change issue effectively. However, the present study has made an attempt to understand scientific issue from the perspective of international politics and international security but precisely focuses on the Climate change frameworks, agreements and reports of the governmental and non-governmental organisations.

NOTES

1. Hergerl, G.C.; Zwiers, F.W.; Braconnot, P.; Gillet, N.P.; Luo, Y.; Marengo Orsini, J.A.; Nicholls, N.; Penner, J.E. and Stott, P.A., Understanding and Attributing Climate Change. In: Climate Change 2007: The Physical Science Basis. Contribution of Working Group I to the Fourth Assessment Report of the Intergovernmental Panel on Climate Change.
2. UNFCCC, Report on Climate Change: Impacts, Vulnerabilities and Adaptation in Developing Countries. Accessed through http://unfccc.int/resource/docs/publications/impacts.pdf on 17 November 2016.
3. Climate Change, Basic Information. Accessed through https://www.epa.gov/climatechange/climate-change-basic-information on 18 November 2016.
4. IPCC, 2013, Summary for Policymakers. In Climate Change 2013: The Physical Science Basis. Contribution of Working Group I to the Fifth Assessment Report of the Intergovernmental Panel on Climate Change [Stocker, T.F., D. Qin, G.-K. Plattner, M. Tignor, S.K. Allen, J. Boschung, A. Nauels, Y. Xia, V. Bex and P.M. Midgley (eds.)]. Cambridge, UK, NY: Cambridge University Press.
5. Thompson, L. (2010), Climate Change: The Evidence and Our Options. *The Behaviour Analyst*, No.2, 33, 153-70
6. Tubi A, Fischhendler I and Feitelson E, 2012, The Effect of Vulnerability on Climate Change Mitigation Policies. *Global Environmental Change*. 22, 472-4-82
7. Barnett J., 2003, Security and Climate Change. *Global Environmental Change*. 13, 7-17
8. Parry M., (2001), Viewpoint: Climate Change-Where Should our Research Priorities be? *Global Environmental Change*. 11, 257-60
9. Klein R.J.T., Schipper E.L.F. and Dessai S 2005 Integrating Mitigation and Adaptation into Climate Change and Development Policy: Three Research Questions. *Environmental Science and Policy*. 8, 579-88
10. Laukkone, J., Blanco, P.K., Lenhart, J., Keiner, M., Cavric, B and Njenga, C.K. 2009 Combining Climate Change Adaptation and Mitigation Measures at the Local Level. *Habitat International*.33, 287-92.
11. Martens P, McEnvoy D and Chang C, (2009), The Climate Change Challenge: Linking Vulnerability, Adaptation and Mitigation. *Current Opinion in Environmental Sustainability*. 1, 14-18.
12. Lorenzoni I, Jordan A, Hulme M, Turner R.K. and O'Riordan T, 2000 A Co-evolutionary Approach to Climate Change Impact Assessment: Part I. Integrating Socio-economic and Climate Change Scenarios. *Global Environmental Change*. 10, 57-68.
13. Tol R.S.J., Downing T.E, Kuik, O.J. and Smith J.B., 2004 Distributional Aspects of Climate Change Impacts. *Global Environmental Change*.14, 259-72.
14. Ikeme, J, 2003 Equity, Environmental Justice and Sustainability: Incomplete

Approaches in Climate Change Politics. *Global Environmental Change*. 13, 195-206.

15. Kaechele, H, Amjath-Babu, T.S., Kutter, T, Specht, K, Nautiyal, S, Muller, K and Raju, K.V., 2011 Confronting Climate Change Challenge: Discussing the Role of Rural India under Cumulative Emission Budget Approach. *Environmental Science and Policy*. 14, 1103-12.
16. Rong F, 2010 Understanding Developing Country Stances on Post-2012 Climate Change Negotiations: Comparative Analysis of Brazil, China, India, Mexico and South Africa. *Energy Policy*, 38, 4582-591.
17. Sathaye J, Shukla P.R. and Ravindranath N.H., 2006 Climate change, sustainable development and India: Global and National Concerns, Current Science, 90 No.3, 314-25.
18. Baettig M.B., Brander S and Imboden D.M., 2008, Measuring Countries' Cooperation within the International Climate Change Regime, Environmental Science & Policy 11, 478-89.

3

Conceptual Framework

3.1 Climate Change

Climate Change is the rise in average temperatures on Earth, mostly due to the burning of fossil fuels. Climate change, also called global warming refers to the rise in average surface temperatures on Earth. An overwhelming scientific consensus maintains that climate change is due primarily to the human use of fossil fuels, which releases carbon dioxide and other greenhouse gases into the air. The gases trap heat within the atmosphere, which have a range of effects on ecosystems, including rising sea levels, severe weather events, and droughts that render landscapes more susceptible to wildfires. Climate Change is a complex phenomenon; consequently, it is also a difficult concept to define. A general definition of this topic is a short-term or long-term alteration of the statistical properties of a climate system. Such a change can be temporary or permanent, which can occur regionally or globally. However, in recent times the main focus is on human activity that is responsible for climate change. Thus the United Nations Framework Convention on Climate Change defines this subject as a "a change of climate which is attributed directly or indirectly to human activity that alters the composition of the global atmosphere and which is in addition to natural climate variability observed over comparable time periods." There is lots of evidence that tells us the average temperatures of the world's atmosphere and oceans have increased over the last 150 years, Such as follows:

- Direct temperature measurement on land.
- Changes in the dates when lakes and rivers freeze and their ice melts.
- A reduction in the extent of snow in the Northern Hemisphere.
- A reduction in glaciers.
- Extended growing seasons of plants.
- Changes in the heat stored in the ocean.
- Changes in rainfall patterns resulting in more floods, droughts and intense rain. Biological changes, like shifts in the ranges of some plant and animal species earlier timing of spring events such as leaf-unloading, bird migration and egg laying for some species, etc.

3.2 Causes of Climate Change

Although scientist, environmentalists, politicians, economists and the general public are highly worried about the consequences of climate change caused by humans, there are also natural factors accountable for this occurrence. Therefore variations in solar radiation, deviations in the Earth's orbit, mountain-building and continental drift and changes in greenhouse gas concentrations are some of the other aspects that must be considered when discussing climate change. Natural factors that cause climate change take a long time (sometimes centuries or millions of years) to produce their impact. For instance the formation of an isthmus in Panama River million years ago prevented the mixing of the water of the Atlantic and Pacific oceans. While, the results of human activity that influence climate change are observable in a relatively short period of time. There are multiple complementary and alternative opinions that attribute the causality of climate change elsewhere. One of the best-supported theories maintains that humans are simply not significant enough and cannot have a profound impact on the global climate system to cause global warming. Thus, something bigger must be going on, such as the sun heating up and warming the entire solar system. A second alternative theory is that warming is based upon heat generated by human waste heat, via the second Law of Thermodynamics, and the order and complexity of modern existence puts off sufficient heat to warm the atmosphere. A third alternative approach makes a variety of assertions, including claims that climate change is not significant enough to be anything more than random fluctuations, that there are still legitimate scientific evidentiary disputes, or that actions to correct warming will cause unacceptable economic damage, in rejecting the greenhouse gas driven explanation.

Over most of the Earth's history, natural processes have been responsible for periods of climate change. Changes in the past can be explained by natural factors such as changes in the Earth's orbit, in the sun's intensity, in the amount of explosive volcanic activity, by changes to the surface of the Earth, and farther back in time, to the position of the continents. Of these, only changes in the sun's intensity and volcanic activity are relevant on century timescales. Human activity has now become the main cause of recent climate change. Global warming is primarily attributed to the enhancement of the natural greenhouse gas effect. Climate change is a warming trend, not just a warming cycle. Global temperature naturally varies from year to year and decade to decade. Natural climate variability will continue to have an influence on the state superimposed on these natural fluctuations as a long-term trend towards global warming.

In order to detect climate change a long-term trend above the 'noise' of natural climate variability, it is important to look to long term-data records. A long -term global warming of about 0.85°C has been observed over the period 1880 to 2012. Climate change will affect communities all over the world. Climate change is projected to lead to both changes in average conditions and in extreme weather events. Climate change is also arguably one of the most complex global policy issues to have arisen so far. The questions involved relate to numerous disciplines, in the pure and applied sciences, positive social science, and political economy, besides ethics and morality. Analysis of the divergent facets of the issue is likely to proceed at the cutting edge of current human knowledge and understanding, and indeed may involve several extensions to the frontiers. As highlighted at the last climate talks in Copenhagen climate change conference (2009), anthropogenic climate change is affecting those who are least responsible for it, mainly the poorest and therefore most vulnerable regions of the world. This risk is growing as industrialised countries delay strong measures.[1] Societies with more climate-sensitive economies, largely in the developing world, will be most affected as climate change acts as a "threat multiplier" and increases existing burdens.[2] There is indeed consensus that nuclear power is a relatively low-cost option providing close to zero-emissions energy and has the capacity to provide significant amounts of energy.

But apart from the high start-up costs, regulatory and decommissioning costs, the far greatest environmental and economic challenge is the disposal of nuclear waste. Long-term storage is very costly and bears radioactive spillage risks. Perhaps reflecting these challenges countries, such as the United Kingdom, who have benefited from nuclear energy, have not commissioned new nuclear power plants.[3] The report from the UN's intergovernmental panel on climate change concluded that climate change was already having effects in real time-melting sea ice and thawing permafrost in the Arctic, killing off coral reefs in the oceans, and leading to heatwaves, heavy rains and mega-disasters.[4] And the worst was yet to come. Climate change posed a threat to global food stocks, and to human security, the blockbuster report said.[5] "Nobody on this planet is going to be untouched by the impacts of climate change," said Rajendra Pachauri, chair of the IPCC.[6]

3.3 Climate Change and Security

According to IPCC 2001, as sovereignty over delineated territory is the material substrata of national security, then physical processes such as sea-level rise may undermine national security in serious ways: for example a 45 cm rise in sea-level will potentially result in a loss of 10.9 per cent of Bangladesh's territory, forcing some 5.5 million people to relocate. Hoegh-Guldberg et al., showed that the complete loss of habitable territory that may occur on small islands as a result of rising seas and increasing climate variability is very severe and the socio-economic impacts of global warming on islands may be "so profound that they dwarf any strategic issue currently confronting a major peacetime economy."[7]

According to Rapkin and Avery, national security has an internal dimension that is partly a function of state legitimacy and the governments for whom the material well-being of their people is highly sensitive to external forces such as changing terms of trade, or where material well-being is in decline, tend to be relatively more unstable, and the country relatively more prone to internal violent conflict.[8] The increasingly frequent and severe hazardous events that are expected as a result of climate change mean more exogenous shocks to all countries, and so possibly less security. Climate change may have many other indirect negative effects that can undermine legitimacy, it may: undermine individual and collective economic livelihood; affect human health through reduced availability of freshwater and food, and by exposing people to new disease vectors; undermine state wealth and military capability; and exacerbate inequalities between people.

The World Bank estimated the losses due to climate change in Kiribati to be in the order of 17-34 per cent of current GDP by 2050. Several measures that are or will be implemented to reduce greenhouse gas emissions could also impose costs to national economies.[9] Goldstone argued that Saudi Arabia already demonstrates demographic and development characteristics conducive to political unrest, a situation which may intensify as a result of implementation of the Kyoto Protocol.[10] Finally, given the dominant concern of national security planners and policymakers with violent conflict and trans-border incursions, the issue of whether, and how, climate change may stimulate and increase migration is important. India's economy and population have been growing steadily since India achieved independence in 1947. Socio-economic development has been one of India's top priorities but India also believes that an assured level of regional stability is very much necessary for this development. To this end, India has traditionally pursued a broad policy of defensive defence.

However military developments, and growing hostilities with Pakistan, have caused her to shift to a strategy of war prevention.[11] As India's growth is involving more and more with economic, technological, and political developments in Asia and the rest of the world, it is also squarely confronted by the consequences of strategic change in the international security order. Developments shaping India's national security environment are occurring globally, in adjacent regions, within the Indian sub-continent, and domestically. The key developments include nuclear proliferation, the growing ascendance of economic power as a key factor in international relations, India's relationships with the United States and Russia, turbulence in Central Asia and Afghanistan, global terrorism, and continuing enmity with China and Pakistan.[12]

Barnett J (2001) gave a heuristic guide to the broader environment and security literature to assess the array of linkages between the particular problem of climate change and security as shown in Figure 10A.

Figure 10(A): Guide to Environment-Security Linkages

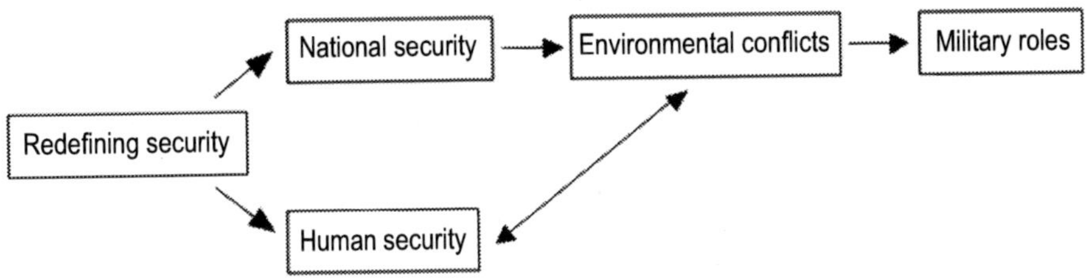

Source: Barnett J: Security and Climate Change, Tyndall Centre for Climate Change Research Tyndall Centre for Climate Change Research Working Paper 7. 2001

The argument that environmental change was a security issue for nations and people was made after 1989 increasingly, in both environmental and security journals that must be considered in the context of dramatically improved relations between NATO and the former USSR, making conventional understandings of security less relevant. Simultaneously, environmental concerns were increasingly coming to the fore of national and international politics and policy.[13] Climate change is a security problem for some states, communities through its likely impacts on ecosystems and their inhabitants. It also has indirect effects on development and political stability. Environmental security was originally written with the intention of exposing the inadequacy of militarized practices of security, the porous nature of sovereignty in the face of environmental change, and to elevate environmental problems from the level of 'low politics' to 'high politics' so that states would commit as much energy and resources to address environmental problems as they do to other security problems.[14] The continent, however, has experienced several major environmental conflicts in the past decades, particularly in the Sahel and Sub-Sahel belt stretching to the Horn of Africa.

The major environmental conflicts which have occurred during the period of 1980-2005 have been shown in Figure 10B.

Figure 10(B): Environmental Conflicts 1980-2005

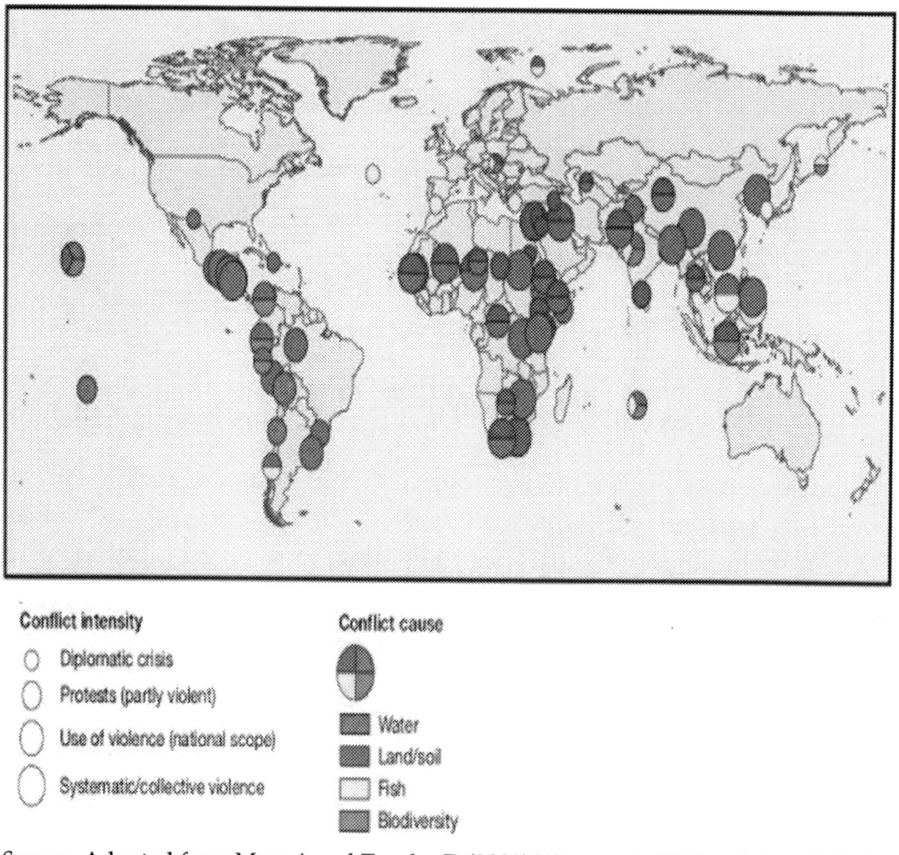

Source: Adapted from Mass A and Tanzler D (2009) Wissenschaftlicher Beirat Globale Umweltverändung (WBGU) 2007 and Carius et al., 2006).

3.4 Security Aspects of Climate Change

Climate change has the potential to be the greatest security threat in this century. It is considered a multiplier of a range of security threats from conflict exacerbation to migration, water food and health security. The focus here is on how the foundations of broadly defined human security are threatened by climate change, as it undermines the pillars of prosperity: water security, energy security, food security and climate security. The main aim is to galvanise greater urgency in mitigating the drivers of climate change and, to a lesser extent, to increase action to adapt to climate change. No one can dispute the overwhelming scientific evidence that indicates climate change as a serious global threat, which demands an urgent global response. What is new is that governments are increasingly treating climate change as a security threat. In an interdependent world, governments are more and more aware that the impacts will be felt not just in the immediate region affected, but also across the international community. Climate change poses many hard security challenges, but there are no hard security solutions. Rather, it is a reflection of the fact that the security sector, which comprises a diverse range of actors (diplomatic, military, intelligence, policing, peace-building, development and humanitarian actors) needs to incorporate climate change into their already complex and challenging agendas. The development community has generally resisted the language of security on the grounds that, as with the post-9/11 rhetoric of the 'War on Terror', the strategic thrust of any security debate will generate short-term and at times confused policy responses. 'Human security' and 'state security' are not mutually exclusive concepts without human security traditional state security cannot be achieved and vice versa. Using the language of human security in relation to climate change stresses that the proper referent for security should be the individual as part of a wider community rather that the state as an entity in and itself.

The polarization between those adversely affected by climate change and those responsible for causing the changes means that climate change is becoming a major obstacle to continued poverty reduction, with all that, that implies for human security. Many developing countries are especially vulnerable to climate change because of their geographic exposure, low incomes and their greater reliance on climate sensitive sectors such as agriculture. The 2007/08 UNDP report on climate change and human development suggests that changed runoff patterns and glacial melt will add to ecological stress, thereby compromising flows of water for irrigation and human settlements in the process. The relationship between food insecurity, malnutrition and health is well documented amongst development experts. Climate change will potentially exacerbate this vulnerability since changes in temperature and precipitation are likely to increase the geographic range of vector-borne diseases such as malaria, cholera and diarrhoea. This is likely to generate higher morbidity and mortality rates among people, especially children, suffering from malnutrition than among food secure people. UNDP estimates that an additional 220-400 million people could be exposed to malaria a disease that already claims 1 million lives annually. Drought and other climate-related shocks risk sparking violence and conflict. Resource-driven conflicts are not new, but in climate change there is a potentially new and deadly dynamic. It is difficult to predict what the human and state security implications would be of such a large displacement. Particularly, Egypt as a case, which is a signatory to the 1951 and 1967 conventions on refugees and has history of being destination country for both traditional refugees and economic migrants largely from bordering countries and Africa, whose status as the country itself is at risk due to major devastation of environmental change.

According to Egypt's Ministry of Environment, there are millions of people who are at risk of forceful migration in the absence of sustainable livelihoods caused by climate change. The coastal zone which is home to more than 40 per cent people, likely to face a severe consequence of climate change as rising of sea level has potential to cause flood and salt water intrusion will compel peoples to abandon their land and homes would create a social instability.[15] In addition, displacement of 2 million people from one of the most fragile regions of the world is bound to have a serious impact and not least on Egypt's internal security and stability. Climate change poses many new threats and challenges to national security and international stability as well as to human security at other scales. The concept of human security was introduced first by UNDP in 1994 and then developed further in a report by the Human Security Commission, co-chaired by Sadako Ogata and Amartya Sen, in its report Human Security Now (2003). The environmental dimension of human security has been addressed by an international team working on Global Environmental Change and Human Security (GECHS) and in several studies by the United Nations University Institute for Environment and Human Security (UNU-EHS).

The impacts of climate change will fuel the politics of resentment between those mostly responsible for climate change and those who are worst affected by it and it will, act as a threat multiplier; deepen political tensions nationally and internationally. Lest the threats posed by climate change are mere futurology, the UN's appeals for emergency humanitarian aid in 2007 were all, bar one, connected to climate change. UNDP will take up the relationship between human development and climate change in its Human Development Report 2007 (HDR), to be launched in November. By recalling the 2007/2008 issue, HDR applied a human development to highlight the costs of climate change on 4 November 2010.[16] Over 17 years, UNDP has incrementally developed a sensitive measure of human development [the Human Development Index (HDI)]. A Global Report Reducing Disaster Risk a challenge for Development, UNDP report published in 2004 have shown that the HDI correlates well with measures of disaster risk such as UNDP's disaster risk index (DRI), especially for less developed countries. Preliminary analysis for this year's (2007) HDR suggests that climate change poses major obstacles to progress in meeting MDGs and maintaining progress raising the HDI: "There is a clear and present danger that climate change will roll back human development for a large section of humanity, undermining international cooperation aimed at achieving the Millennium Development Goals (MDGs) in the process."[17]

- If temperature continues to rise, we can expect an additional 393,000 deaths by the year 2030 in sub-Saharan Africa, argued Marshall B. Burke and others in an article in proceedings of the United States National Academy of Science published in 2009. Such alarmist claims arouse interest among international organizations and national agencies dealing with security issues, ranging from national governments and their intelligence communities to international bodies, such as the United Nation Security Council.
- But can such contentions be substantiated?
- There is no doubt that humans are driving a gradual warming of the planet and are responsible for changes in hydrological cycles. Due to modern satellite imaginary, a 3.1 cm sea-level rise has been observed between 1993 and 2003. In most-regions of

the world, the intensity to tropical cyclones has increased. Heat waves in Europe have become more frequent. In 2019, global mean sea level was 3.4 inches (87.61 mm centimetres) above the 1993 average—the highest annual average in the satellite record (1993-present). From 2018 to 2019, global sea level rose 0.24 inches (6.1 millimetres).

Climate change is a fact. But do these changes drive armed clashes? Apart from Burk's study, there is still little empirical evidence that Climate change will lead to more conflicts. However, these changes will most certainly have serious implications for human well-being. These linkages are complex in many ways. To begin with, climate change involves the interactions of many systems such as the atmosphere, hydrosphere, cryosphere, and biosphere that are immensely complex in their own right. Thus, a recurrent theme in IPCC reports is the significance of thresholds and nonlinearities. When human systems are added to the mix, complexity escalates. Livelihood security and other aspects of human security interact with geo-strategic (or "hard") security issues because of the national and regional upheavals that climate stress may put on livelihood systems already vulnerable and incapable of adapting. The rural and urban poor worldwide are already under stress, and for some groups such as women headed households in Africa, adapting on to climate-induced stress will be very difficult indeed.

As Simon points out, climate change has both intermittent but increasingly frequent, extreme impacts (such as severe storms and heat waves) and slow onset, pervasive, cumulative effects (such as extinction of life forms and sea level rise). Both kinds of effects may have a role in displacing human populations and disrupting their livelihoods. Some major climate changes may actually occur rapidly.[18]

3.5 Security Implications of Climate Change

There are at least seven ways that climate change is likely to affect security in its narrow and wider meanings:

3.5.1 Food-Livelihood-Climate-Conflict

Food and livelihood pressure due to climate change will lead to populist and/or military coups in a number of countries. The same pressures as well as the "push" provided by conflict will cause considerable population movements and displacement both within countries and internationally. That, in turn, will increase insecurity a process that is already occurring.[19]

3.5.2 Health-Climate-Livelihood-Conflict-Security

Health impacts of climate change including epidemics and insect outbreaks, will have a similar effect, also compounding food and livelihood crises. This is beginning, but it will accelerate in the medium term. As we have seen with SARS and avian influenza, disease does not respect national boundaries in a globalized world. Changing climate may bring many epidemiological surprises as vector habitats change, sometimes quite rapidly. The recent outburst of COVID-19 is a larger example of it which respects no boundaries and created havoc globally by causing health crisis and economic slowdown due to worldwide lockdown. Undoubtedly, COVID-19 has affected millions of lives including the environment under which the CO_2 emissions dropped down as industries along with other means of transports were stopped helped to improve air quality yet its sustainable impact on the environment for longer-term remained a major concern.[20] Projected climate change-related exposures are likely to affect the health status of millions of people, particularly those with low adaptive capacity, through:

- increase in malnutrition and consequent disorders, with implications for child growth and development;
- increased deaths, disease and injury due to heat-waves, floods, storms, fires and droughts;
- the increased burden of diarrhoea;
- The increased frequency of cardio-respiratory diseases due to higher concentrations of ground level ozone related to climate change; and, the altered spatial distribution of some infectious disease vectors.[21]

3.5.3 Mitigation-Adaptation-Inequity-Conflict-Security

The social and economic impacts of climate change focused mitigation projects have not yet been fully assessed. Skutsch, for example, highlights the absence of gender impact analysis of CDM projects. The "clean, green" solution of nuclear energy so heavily promoted in the early years of this century and recently adopted as a climate change-fighting strategy by the EU and others has led to a planet-wide race to capture extractive rights and exploit uranium reserves. The historically high value of uranium has led to violent conflict in the Congo and renewed the use of forced labour. The profound, degenerative human health consequences of working and living in a toxic and radioactive environment are only recently being understood. Cases of cancer and other radiogenic health problems especially in the areas, hosted the Cold War nuclear machine. In addition, a new phase in the growth of nuclear power could exacerbate the problem of "leakage" of weapons grade material into the hands of non-state actors.[22]

3.5.4 Food-Fuel Conflict and Human Security

Large-scale investments in biofuels in the medium and long term as a substitute for greenhouse gas-producing petro-based energy sources may have its own effect of taking considerable land out of food production and diverting food grains, thus raising food prices and eroding biodiversity.[23]

3.5.5 Disaster Livelihood Governance Conflict Security

Climate change is likely to shape the developmental trajectory for the upcoming future by influencing the nature, frequency and scale of humanitarian crisis.[24] The extreme weather events become more intense and/or more frequent; the economic and social costs of those events have increased. The poor countries like Bangladesh in Asia and Madagaskar in Africa are suffering due to deteriorating climatic conditions. Climate change impacts spread from directly impacted areas and sectors to other areas and sectors through extensive and complex linkages. Disasters, development, and conflict have been shown to have complex interactions with one another quite apart from the additional stresses and management challenges likely to accompany increasingly variable weather and more extreme storms.[25]

3.5.6 Water Conflict

Much discussed (incorrectly) in the past as a population growth/

demand problem, water scarcity may, indeed, become a source of conflict and stability, and more so as a function of supply. Water and climate conditions in the world are expected to change immensely due to population growth, economic development and climate change. It is expected that these developments will lead to increased water stress in the future and risks of conflicts worldwide. At greater extent, water-related risks undermine human well-being and contribute to political instability, violent conflict, human displacement, migration, and acute food insecurity. Further, it also undermines national, regional, and even global security. The water risks have threatened human civilizations over millennia.[26] In terms of human security, throughout history, water has played an important role. But, attention to the water-related issues and threats have been growing in recent periods due to increasing water risks since the population growth and economic expansion have increased demand for water, distinctly in various regions of the world.[27]

In future, water demand is going to rise, much higher than today due to increase in population and economies projected to grow to nine billion in 2050.[28] Water crises are increasing around the world and will be worsened by climate change. Water has often overlooked as a root cause of conflict and destabilizing migration. For instance, countries like Iraq, Iran, Mali, Pakistan, India and Nigeria; experienced siring up tension over water includes protests and violence. The water shortage scarcity of land resources, poor quality of drinking water remained major source of conflict and other factors such as ethnic diversities and presence of external armed groups deteriorated the situation.[29] Also, when shifts in water security combined with the degradation of wetlands, competition for natural resources such as fisheries and pasture intensified that can escalate into violence and displacement, especially in fragile states.[30]

Today, water insecurity is a growing threat around the world, and non-addressing will get worse in a void of immediate innovative solutions. Besides, Climate change is creating an extraordinary challenge and urgency to solve water issues. But, non-profit organizations like The Nature Conservancy is unable to address individually, investment in technology and breakthrough collaborations are necessary to make significant progress and ensure a water-secure future and prevent conflict.[31] The other tensions and sources of instability discussed above could also place international water management and sharing agreements under pressure.[32]

3.5.7 Sea Level Displacement Security

Finally, sea level rise in the long term and its collateral impacts on river flow and ocean discharge regimes will cause displacement of millions of people currently living in coastal areas. For the small island nations of the world, especially the many cultural groups living on coral atolls, entire nations face complete submersion A recent study calculates that around 634 million people are living less than 10m above sea level. The study notes "that of more than 180 countries with populations in the low-elevation coastal zone, about 70 percent have urban areas of more than five million people that extend into it."[33] We know that climate change is truly a global challenge with no part of the world being spared from the consequences. Even if we stabilise the flow of CO_2 to the atmosphere we cannot avoid an increasing stock of CO_2. Adaptation will become necessary, as the only solution for many developing countries. The solutions need the involvement of all segments of society.

Consumer preferences are important, from shifting eating habits and the consumption of less meat to energy-efficient household appliances. There are proven technologies, like hydropower with known and limited unintended effects. And most important, we need to understand better that we are both agents and victims of climate change, in all regions of the world. If we do not act now, the consequences will not only be very costly, but will also become a serious threat to human mankind in addition to infectious diseases, the threat of nuclear war and terrorism. What is the biggest national security threat? Is Climate change the biggest national security threat? These questions confuse the nature of today's security threats, and more specifically, obscure the complex way in which climate change affects the border security landscape. Climate change is not an exogenous threat, hermetically sealed from other risks. It is, as the CAN (Channel News Asia) corporation first stated in 2007, a "threat multiplier". The impacts of climate change interact with other factors to make existing security risks, whether it's state fragility in the Middle East, or territorial disputes in the South China Sea-worse.

Furthermore, there's a tendency to think of the security implications of climate change in a conflict context only. But this is for too narrow Climate change intensifies natural resource stresses in a way that can increase the likelihood of livelihood devastation, state fragility, human displacement and mass death. These dynamics do not always result in conflict, but they certainly represent a threat to human, national, regional, and in the right context, international security so if we narrow the discourses too much, and focus only on conflicts, we run risk of being unprepared for a range of possible scenarios. In short, we need to move away from both 'ranking' threats to national security or focusing on just one element of risk landscape that's easily understood. We need to take a closer look at how security risks are connected, and build from there. For example: What will a rapidly melting Arctic mean for US-Russia relations? Will increased water scarcity in North Korea or Iran have implications for higher–order security issues? Will competition over migrating fish stocks increase interstate tensions in a warming South China Sea? In other words, the question is not do we deal with climate change, or deal with Russia, ISIS and Iran because; they cannot be so easily delinked. The pertinent questions are: How does climate change affect our relations with Russia? How does it exacerbate instability in the Middle East? How might it help create stressed environment that non-state actors can more easily exploit?

3.6 Direct and Indirect Threats?

Climate change is a "threat multiplier", exacerbating existing risks to security. The threat comes not from climate change itself, but rather, from how it interacts with existing security conditions. In this context, climate change presents both direct and indirect threats to human, national and international security.

3.6.1 Direct Threat

Climate change has a direct impact on security through its effect on the critical infrastructure underpinning a nation's security. This includes sea level rise risks to military installations that can degrade a nation's ability to conduct military operations, as well as extreme weather events that can devastate essential energy, financial and agricultural centres that undergird a nation's economic viability. In some cases, as with some low-lying small island states, sea level rise presents an existential threat.

3.6.2 Indirect Threat

Climate change also presents an indirect threat to security by increasing stresses on the critical resources underpinning a nation's security, including water, food and energy. These stresses can degrade a nation's capacity to govern. Decreases in water, food and energy availability can devastate livelihoods, and contribute to a broad range of destabilizing trends, including internal population displacements and migrations, and political unrest. These pressures in term can contribute to state fragility, internal conflict and eventually state collapse. Climate change can also indirectly change or disrupt existing international security dynamics in geostrategic environments, such as the Arctic and the South China Sea.

Climate change poses significant security risks and grave challenges to the Asia-Pacific region, including: decreasing energy access, decreasing access to food, increased frequency and intensity of hydro meteorological disasters, population displacement, increases public health problems and water stress. The kinds of responses that could help manage these climate security risks include policy and institutional strengthening; regional and international cooperation, monitoring, research and technology, public awareness, education and training and finance and resource mobilization. It would be prudent to begin implementing these responses in the near term, recognizing that there is uncertainty about the timing of these security risks, some of which will emerge gradually and some of which may be a rapid onset and catastrophic in nature. The security issue is of increasing concern and is therefore consistent with the United Nations Framework Convention on Climate Change and the Kyoto Protocol's objective to avoid dangerous climate change.

The UNFCCC entered into force on 21 March 1994. The ultimate goal of the UNFCCC is to preventing "dangerous" human interference with the climate system.[34] On the other hand, the Kyoto Protocol was adopted on 11 December 1997, owing to a complex ratification process, entered into force on 16 February 2005. Currently, there are 192 Parties are involved to the Kyoto Protocol. The Kyoto Protocol operationalizes the United Nations Framework Convention on Climate Change by committing industrialized countries and economies in transition to limit and reduce greenhouse gases (GHG) emissions in accordance with agreed individual targets.[35]

3.7 Contemporary Problem of Global Peace and Security

Climate change like many global problems now recognized as a threat to the international security and cooperation. In theoretical terms, it is being securitized and included in the traditional security studies. Climate change and its accompanying environmental degradation are perceived to be threats that can have incalculable consequence on the international community. The consequences are said to have more effects in small island developing nations. In recent years, the security implications of the climate change are being addressed at the national, regional and multilateral levels. Global climate change challenges have been at the forefront of the multilateral agenda in the last quarter of a century. Beginning with the Montreal protocol on substances that deplete the Ozone Layer (Montreal Protocol) of 1987, and the Rio Conventions (UN Convention on Biological Diversity, UN Framework Convention on Climate Change, and UN Convention Combat Desertification) of 1992, Climate change issues have become policy area concerns for many governments across the globe. As such, because of the nature of the global climate change challenges, the international community through the United Nations Framework Convention on Climate change (UNFCCC) had embarked on comprehensive negotiations aimed at finding adequate solutions for mitigating global climate change.

Ever since the Kyoto Agreement came about, the global climate change politics has been dealing with the question of who should pay the most and what responsibility should the traditional emitters and largest emerging developing countries bear in financing the effects of the climate change in countries and regions of the world. Over the last two decades or so, there has been growing concern from the international community over possible security implications of the global climate change. The United Nations (UN), European Union (EU), United States (US), African Union and many other regions of the world have recently begun to ask how the security dimensions of climate change will affect-access to shared resources, national security, borders integrity, trade routes and international political alliances and economic relations.

By maintaining international peace and security, the UN makes all its work more effective, because by keeping the peace, the organization can focus on solving global issues, instead of resolving conflicts. Unfortunately, recent peace and security challenges in areas where the organization has a limited presence have tested its ability to maintain peace. In areas where the UN has a presence, it has increasingly come under attack. On 9 January 2004, David King, the UK Government's chief scientific adviser claimed that climate change is a far greater threat to the world than international terrorism. In February 2004, John Reid MP, then British Secretary of State for Defence and now Home Secretary, argued that climate change may spark conflict between nations. He forecasted that violence and political conflict would become more prevalent in the next 20 to 30 years as climate change turned land into desert, melted ice fields and poisoned water supplies. He listed climate change alongside the major threats in future decades, including terrorism, demographic changes, and global energy demand. "As we look beyond the next decade, we see uncertainty growing; uncertainty about the geopolitical and human consequences of climate change. Impacts such as flooding, melting permafrost and desertification could lead to loss of agricultural land, poisoning of water supplies and destruction of economic infrastructure. More than 300 million people in Africa currently lack access to safe water; climate change will worsen this dire situation". John Ashton, the UK Foreign Secretary's Special Representative for Climate Change, said at a conference on "Climate Change: The Global Security Impact", at the Royal United Services Institute on 24 January 2007: "There is every reason to believe that as the 21st century unfolds, the security story will be bound together with climate change." "Climate change is a security issue because if we don't deal with it, people will die and states will fail," Ashton concluded.[36]

In June 2019, Prime Minister Theresa May said there was a "moral duty to leave this world in a better condition than what we inherited". According to her, cutting emissions would benefit public health and also cut the National Health Service costs.[37] The UK has set a target in 2050 to reduce emissions by 80 per cent under the Climate Change Act of 2008.[38] But, now it will amend to the new, as it seems tougher to achieve such target. However, the actual terminology used by the government is "net zero" greenhouse gases by 2050.[39] By doing so, Britain has become the first major nation to propose this target. How will changing climate patterns affect both inter-state relations and international and national security in a narrow, geo-strategic sense as well as the well-being and survival of human beings and, water, health and livelihood security?[40] On 6 April 2007, the Intergovernmental Panel on Climate Change (IPCC) released the second part of its Fourth Assessment Report showing that the poor of this planet are most likely to suffer the worst effects of climate change. Human security takes on a broader meaning when one considers basic needs for food, water, health in short, livelihood and a place to live the issues addressed in the Millennium Development Goals.[41] While natural climate variations have existed for millennia, anthropogenic climate change has gradually emerged since the industrial revolution and especially after World War II due to the availability of cheap fossil fuels (coal, oil, natural gas) and the dramatic increase in its consumption first primarily in the industrialized countries but now increasingly also in the rapidly growing economies of the BRIC states (Brazil, Russia, India, China), especially of China and India.[42]

The UN works to maintain international peace and security in a world where security threats have become more complex. Although the organization has had many successes, there are also, unfortunately, several recent tragic cases where the United Nations has not been so successful. In July, the council will hold an open debate on the impact of climate change on the maintenance of international peace and security. The executive director of the UN Environment Programme (UNEP), Achim Steiner, is expected to address the council, and the Secretary General may attend. Many countries are also expected to participate in the open debate. On 17 April 2007 the council held a high-level debate on the relationship between energy, security and climate, initiated by the UK, which held the council presidency that month. It was the first time the council had addressed the potential impact of climate change on security.

Since 2007, global concerns about climate change and its implications intensified. The Intergovernmental Panel on Climate Change (IPCC) established by UNEP and the World Meteorological Organization (WMO) issued its assessment report, "Climate Change 2007" in November 2007, which highlighted the reality of climate change, its impact and possible consequences. On 3 June 2009, on the initiative of the small-island developing states of the Pacific Ocean, the General Assembly held a debate on climate change and its possible security implications. On 11 September, 2009, the UN Secretary-General presented his report to the General Assembly. The report identified climate change as a "threat multiplier" that exacerbates existing threat, such as persistent poverty, week institutions for resource management and mistrust between communities. It identified five further ways in which climate change might affect security, which are as follows:

- Climate change could threaten food security and human health and increase exposure to extreme events;
- It could undermine the stability of states by slowing or reversing development;
- It could increase the likelihood of domestic conflicts due to migration and depleting resources, with possible international ramifications;
- Disappearance of territory might aggravate issues of sovereignty, rights and security;
- International conflict might be a result of climate change's impact or demarcated international resources.

In December 2009, the 15th session of the conference of parties (COP) to the UN framework convention on climate change was held in Copenhagen. The 16th session of the COP took place in Cancun, ending in December, 2010 with an agreement that set up a new fund to help poor countries adapt to climate change, and created a new mechanism for transfer of clean energy technology, but again short of specific legal obligations to achieve these ends. By maintaining international peace and security, the UN makes all its work more effective resolving conflicts. The UN works to maintain international peace and security in a world where security threats have become more complex. The line has become increasingly blurred between criminals and hostile groups and peace strategies and sophisticated tactics.

The breakdown of the state security apparatus in intrastate and inter-communal conflicts now poses tremendous security challenges and tests the Organization's capacity to carry out its mandates and programs. Ensuring that United Nations troops are properly supported and equipped is a high priority. The complexity of contemporary peacekeeping environments requires strengthened partnership with all stakeholders, including regional and sub-regional organizations, the wider United Nations family, international and regional financial institutions and donors, and multilateral and bilateral partners. Only through such collaboration can we collectively address the international peace and security: challenges we face now, and in the coming years.

3.8 Climate Change and Security Discourse

3.8.1 Concept of Security

According to United Nations Report on Concept of Security published in 1986, in principle, security is a condition in which states consider that there is no danger of military attack, political pressure or economic coercion, so that they are able to pursue freely their own development and progress. International security is the result and the sum of the security of each and every state member of the international community. However, security is a relative rather than an absolute term. National and international security needs to be viewed as matters of degree. According to a Report published under European Commission Sixth Framework Programme, the first signs of a trend towards the expansion of the notion of security can be traced back to the late 1960s, when Robert McNamara suggested that security implied the freedom of a state to develop and improve its position in the future.

According to Mesjasz C.,[43] a simple scheme as shown in figure 1 can be used to describe that all directions of development of the concept of security provide a ground for prediction (identification) of threats, and planning and preparations of actions, which in turn, are the main goals of any policy-oriented investigations into security.

Mesjasz C.[44] also demonstrated a conceptual framework to show security as a property of social systems as given in Figure 11. The "core

Figure 11: Core Concept of Security (Mesjasz C, 2004).

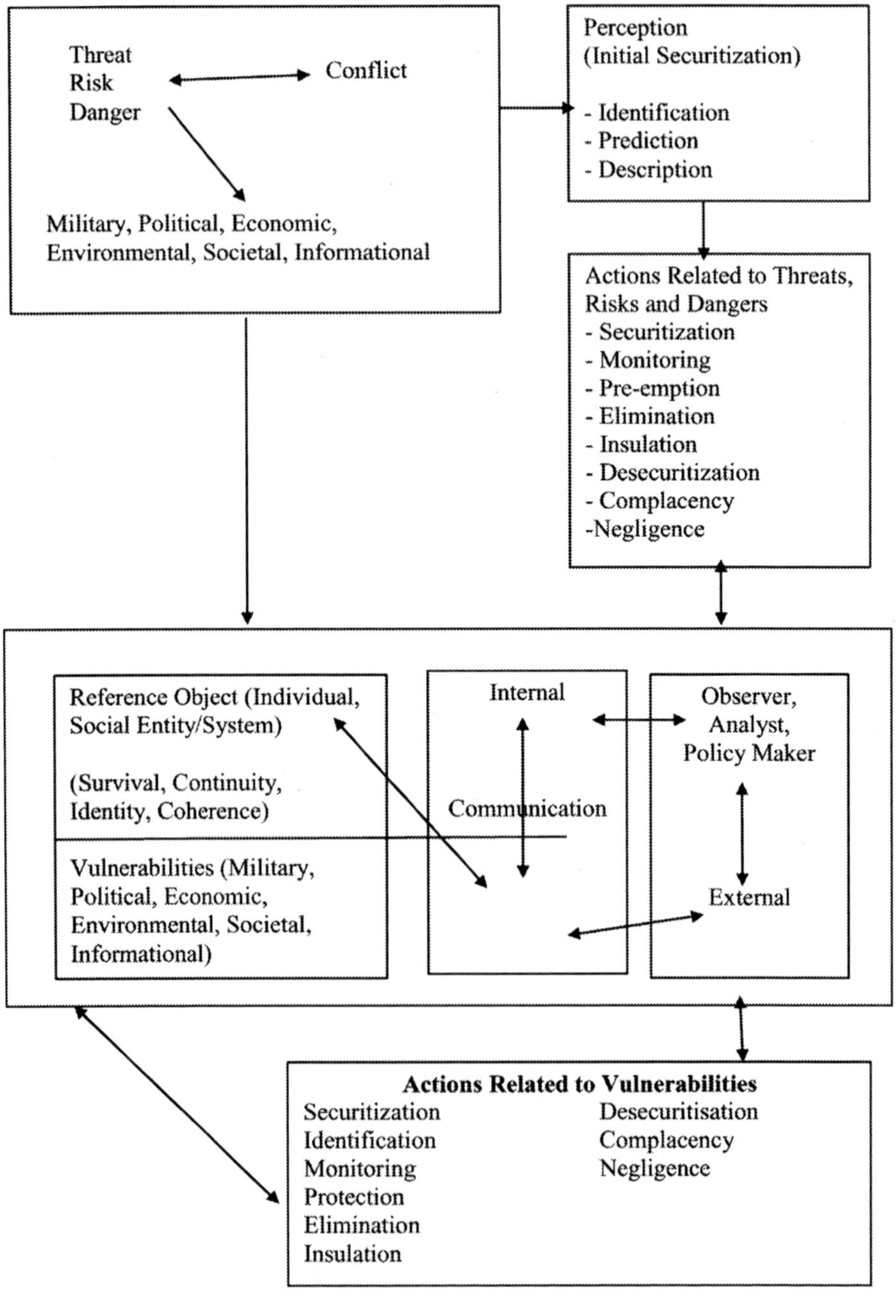

scheme" of security can be extended in various directions by combination of the following attributes:

1. Reference objects like—state, region, alliance, society, various social groups, like—nations, minorities, ethnic groups, individuals, global system.
2. Areas in which existential disturbances (threats) are emerging (sectors)—political, military, economic, ecological, societal.
3. Methods of prediction (identification) of disruptions, beginning from search for "objective" threats and ending with subjectively perceived threats, also resulting from social discourse ("securitization").

Methods of planning and accomplishing extraordinary actions aimed at monitoring, preventing or eliminating existential threats.

Security is an elusive term, which is employed in a wide range of contexts and to multiple purposes by individuals, corporations, governments and academics according to Baldwin D.A.[45] In 1973, Klaus Knorr began a survey of the field by stating his intention to deliberately bypass the semantic and definitional problems generated by the term "National Security". In 1975, Richard Smoke observed that the field had 'paid quite inadequate attention to the range of meanings of 'security' and in 1991, Buzan described security as 'an underdeveloped concept' and noted the lack of 'conceptual literature on security' prior to the 1980s.

The balance of power concept has often been the basis for the formation of military alliances. The balance of power theory is a core tenet of both classical and neorealist theory and seeks to explain alliance formation. It may refer to a situation in which equivalent power is held by two or more nations or groups of nations and to a policy of promoting the creation or preservation of such equivalence in power. According to Kenneth Waltz, founder of neorealism, balance-of-power politics prevail wherever two and only two requirements are met: that the order be anarchic and that it be populated by units wishing to survive. The basic notion of the concept of balance of power in international relations is 'the relationship between the number of actors and the stability of the system'. Deterrence is a security concept whose objective is to dissuade a potential adversary from initiating war, by threatening the use of force in order either to deny an adversary from gaining his objectives by military means or to punish the adversary should he seek to do so.

3.8.2 Deterrence

Deterrence can be divided into three separate categories

(a) **Specific deterrence** focuses on the individual in question. The aim of these punishments is to discourage the criminal from future criminal acts by instilling an understanding of the consequences.

(b) **General or indirect deterrence** focuses on general prevention of crime by making examples of specific deviants. The individual actor is not the focus of the attempt at behavioural change, but rather receives punishment in public view in order to deter other individuals from deviance in the future.

(c) **Incapacitation** is considered by some to be a subset of specific deterrence. Incapacitation aims to prevent future crimes not by rehabilitating the individual but rather from taking away his ability to commit such acts.

Peaceful coexistence was a theory developed and applied by the Soviet Union at various points during the Cold War. This theory was developed in the context of its Marxist– Leninist foreign policy and was adopted by Soviet-influenced "Socialist states" that they could peacefully coexist with the capitalist bloc (i.e. non-socialist states). The Five Principles (set of principles to govern relations between states) of Peaceful Coexistence, known in India as the Panchsheel Treaty (from Sanskritpanch: five, sheel: virtues) are as follows:

1. Mutual respect for each other's territorial integrity and sovereignty
2. Mutual non-aggression
3. Mutual non-interference in each other's internal affairs
4. Equality and mutual benefit
5. Peaceful co-existence

3.9. Measures of National Security

Measures taken to ensure national security generally includes the following measures:

- Using diplomacy to rally allies and isolate threats
- Marshalling economic power to facilitate or compel cooperation
- Using intelligence services to detect and defeat or avoid threats and espionage, and to protect classified information
- Maintaining effective armed forces
- Implementing civil defence and emergency preparedness measures (including anti-terrorism legislation)
- Ensuring the resilience and redundancy of critical infrastructure
- Using counterintelligence services or secret police to protect the nation from internal threats.
- The analysis of the natural environment as an independent and causal variable is largely absent in the field of international relations, which is defined as a branch of political science.

Kenneth Waltz has sealed the ouster of natural environment from the realm of social science as found in his "three images"[46] typology of theories of the causes of war. The first image depicts the location of the causes of international outcomes; say the cause of war, "in the nature and behaviour of men, wars result from selfishness, from misdirected aggressive impulses, from stupidity."[47] The second image identifies the causes in the internal structure of the state. Imperialism for example, results from a particular internal economic structure like capitalism; similarly, international peace results from a particular form of government like democracy.[48] In order to describe actor's behaviour in international realm it engages with other actors under certain norms. These norms are governed under structural conditions. The third image locates causes within the system.[49] According to MacFarlane et al,[50] the origin of the modern concept of "national security" as a philosophy of maintaining a stable nation state can be traced to the Peace of Westphalia, wherein the concept of a sovereign state, ruled by a sovereign, became the basis of a new international order of nation states.

Thomas Hobbes stated that citizens yield to a powerful sovereign who in turn promises an end to civil and religious war, and to bring forth a lasting peace, and give him the right to conduct policy, including wage war or negotiate for peace for the good of the "commonwealth", i.e. a mandate for national security. The Clausewitzian view of diplomacy and war being the instruments of furthering national cause, added to the view of national security being sought by nations by exercising self-interest at all times. This view came to be known as "classical realism" in international relations.[51] National security is particularly potent because the state claims to be the highest form of political organization, and because states possess both much of the legitimacy and much of the capacity to act in the name of security. Security is the key function of the state: because the state is foundational to the provision of order, it has its own claim to a right of survival and self-defence, which is different from the sum of the individual rights to security of its citizens. When the state is both referent object and principal provider of security, this gives a strong territorial and state-centric framing to both the definition of threats and the provision of security. According to Buzan, the security of humans is affected by factors in five major sectors namely:

1. Military security concerns the two-level interplay of the armed offensive and defensive capabilities of states and states perceptions of each other's intentions.
2. Political security concerns the organizational stability of states, systems of government and the ideologies that give them legitimacy.
3. Economic security concerns access to the resources, finance and markets necessary to sustain the acceptable levels of welfare and state powers.
4. Societal security concerns the sustainability, within acceptable conditions for evolution of traditional patterns of language, culture, religious and national identity and custom.
5. Environmental security concerns the maintenance of the local and the planetary biosphere as the essential support system on which all other human enterprises depend.

3.9.1 UN Climate Action Summit 2019

With global emissions reaching record levels and showing no sign of peaking, UN Secretary-General António Guterres called on all leaders for the UN Climate Action Summit 2019 in New York on 23 September. He called for concrete, realistic plans to enhance their nationally determined contributions by 2020, in line with reducing greenhouse gas emissions by 45 per cent over the next decade, and to net zero emissions by 2050. The Climate Action Summit, and the Climate Youth Summit, succeeded in focusing the attention of world leaders, from government, the private sector and civil society, on the urgency for action to address the climate emergency, and on increasing climate action. The Climate Action Summit reinforced the global understanding that 1.5°C is the socially, economically, politically and scientifically safe limit to global warming by the end of this century, and to achieve this, the world needs to work to achieve net zero emissions by 2050. The summit also demonstrated the need to urgently update and enhance their short-term commitments by 2020, and the mid-term commitments by 2030, that will be captured in their national climate plans, known as Nationally Determined Contributions to the Paris Agreement. There was also widespread concern that the world is presently way off course to meet the global target, as emissions continue to increase, and global temperatures rise. The last four years were the four hottest on record, and winter temperatures in the Arctic have risen by 3°C since 1990. Sea levels are rising, coral reefs are dying, and we are starting to see the life-threatening impact of climate change on health, through air pollution, heatwaves and risks to food security. The Summit featured the full participation of governments, business leaders, sub-national actors, indigenous peoples, youth and other civil society stakeholders, who, together, was critical to raising ambition for climate mitigation and adaptation. Together, they launched transformative initiatives in 12 critical areas that will provide the foundation for action going forward to reduce emissions and strengthen adaptation and resilience. The Summit acknowledged the need for pledges to be accountable and provided a baseline for monitoring the progress of the initiatives to ensure that promised results are achieved.

3.9.2 Agenda

The Summit demonstrated the political leadership of 70 countries which committed to deliver more ambitious national climate plans in 2020 in line with net zero emissions by 2050 strategies. While these countries represent a significant portion of the world's population, they account for less than 10 percent of the world's GHG emissions. Some countries were unable to announce commitments at the Summit itself "committed to committing" to new plans by the end of 2020. Finally, 75 countries committed to deliver 2050 net zero emissions strategies by 2020. Through concrete examples, the Summit highlighted how climate action can have tangible and beneficial impacts on people's lives, including on their jobs and health, and therefore the need to align policies and systems to accelerate the implementation of both the Paris Agreement and the Sustainable Development Goals (SDGs). Many businesses showed that they are moving to take climate action. A group of the world's largest asset-owners responsible for directing more than \$2 trillion in investments committed to move to carbon-neutral investment portfolios by 2050, and 87 major companies with a combined market capitalization of over US\$ 2.3 trillion pledged to reduce emissions and align their businesses with what scientists say is needed to limit the worst impacts of climate change a 1.5°C future. In addition, 130 banks one-third of the global banking sector signed up to align their businesses with the Paris agreement goals. The Climate Action Summit set a clear direction for further climate action. Initiatives launched in 12 areas, provide the foundation for scaled-up action. But the Summit also highlighted areas where much more action is needed to secure our footing on a path towards 1.5°C by the end of the century.

3.9.3 UN Climate Change Conference 2019 Highlights

Major announcements by government and private sector leaders at the United Nations Climate Action Summit boosted climate action momentum, and demonstrated growing recognition that the pace of climate action must be rapidly accelerated. Sixty five countries and major sub-national economies such as California committed to cut greenhouse gas emissions to net zero by 2050, while 70 countries announced they will either boost their national action plans by 2020 or have started the process for the same. Over 100 business leaders delivered concrete actions to align with the Paris Agreement targets, and speed up the transition from the grey to green economy, including asset-owners holding over $2 trillion in assets and leading companies with combined value also over $2 trillion. Many smaller countries, including Small Island Developing States and Least Developed Countries, were among those who made the biggest pledges, despite the fact that they have contributed the least to the problem. UN Secretary-General António Guterres, in closing the Summit, said, "You have delivered a boost in momentum, cooperation and ambition. But we have a long way to go." "We need more concrete plans, more ambition from more countries and more businesses. We need all financial institutions, public and private, to choose, once and for all, the green economy." Youth leaders including Greta Thunberg drove home the urgency of greater action by leaders, and their determination to hold leaders to account.

- France announced that it would not enter into any trade agreement with countries that have policies counter to the Paris Agreement.
- Germany committed to carbon neutrality by 2050.
- Twelve countries today made financial commitments to the Green Climate Fund, the official financial mechanism to assist developing countries in adaptation and mitigation practices to counter climate change. This is in addition to recent announcements from Norway, Germany, France and the United Kingdom who have recently doubled their present contributions.
- The United Kingdom today made a major additional contribution, doubling its overall international climate finance to L11.6 billion for the period from 2020 to 2025
- India pledged to increase renewable energy capacity to 175 GW by 2022 and committed to further increase to 450 GW, and announced that 80 countries have joined the International Solar Alliance.
- China said it would pursue a path of high quality growth and low-carbon development, and announced a partnership that could potentially unlock up to 12 billion tons of global emissions

reductions and removals annually through nature-based solutions.

- The European Union announced at least 25 per cent of the next EU budget will be devoted to climate-related activities.
- The Russian Federation announced that they will ratify the Paris Agreement, bringing the total number of countries that have joined the Agreement to 187.
- Pakistan said it would plant more than 10 billion trees over the next five years

3.9.4 On Unprecedented Levels of Private Sector Action

- A group of the world's largest asset-owners responsible for directing more than $2 trillion in investments committed to move to carbon-neutral investment portfolios by 2050.
- Eighty-seven major companies with a combined market capitalization of over US$ 2.3 trillion pledged to reduce emissions and align their businesses with what scientists say is needed to limit the worst impacts of climate change a 1.5°C future.
- Hundred and thirty banks one-third of the global banking sector signed up to align their businesses with the Paris agreement goals.

3.9.5 On Transitioning from Brown to Green Energy

- Michael Bloomberg will increase the funding and geographic spread of his coal phase out efforts to 30 countries. Already, his work has helped to close 297 out of 530 coal plants in the US.
- Countries, including France and New Zealand, announced that they will not allow oil or gas exploration on their lands or off-shore waters.
- Heads of State from Finland, Germany, Greece, Hungary, Ireland, Italy, Netherlands, Portugal, and Slovakia, are among those that announced that they will work to phase out coal. The Republic of Korea announced it would shut down four coal-fired power plants, and six more will be closed by 2022, as well as the

doubling of its contribution to the Green Climate Fund.

- The Summit also delivered critical platforms for improving energy efficiency and reducing the growing energy needs for cooling, with the "Three Percent Club" coalition working to drive a three percent annual global increase in energy efficiency, and the Cool Coalition setting ambitious national cooling targets for its members with the potential to deliver up to 1 degree on the pathway to a 2050 net zero carbon world.

3.9.6 On Scaling up Financing and Unlocking Barriers to Funds

- Many countries announced new contributions to the Green Climate Fund, the official financial mechanism to assist developing countries in adaptation and mitigation practices to counter climate change, with several countries, including France, Germany, Norway and the United Kingdom, announcing that they would double their present contributions.
- Further, the Climate Investment Platform was officially announced today. It will seek to directly mobilize US$ 1 trillion in clean energy investment by 2025 in 20 Least Developed Countries in its first year.
- Summit initiatives were designed to ensure that the actions undertaken would be fair for all, supporting jobs and clear air for better health, and protect the most vulnerable, as well as new initiatives on adaptation, agriculture and early warning systems that will protect 500 million additional people against the impacts of climate change.

NOTES

1. What Will Climate Change Mean for Human Security?, Accessed through https://ourworld.unu.edu/en/what-will-climate-change-mean-for-human-securityon 10 November 2016.
2. Ibid.
3. Ibid.
4. Climate change a threat to security, food and humankind -IPCC report, Accessed through https://www.theguardian.com/environment/2014/mar/31/climate-change-threat-food-security-humankind on 10 November 2016.
5. Ibid.
6. Ibid.

7. Hoegh-Guldberg, O., Hoegh-Guldberg, H., Stout, D., Cesar, H., Timmerman, A., Pacific in Peril: Biological, Economic and Social Impacts of Climate Change on Pacific Coral Reefs. Greenpeace, Amsterdam, 2000.
8. Rapkin, D. and Avery, W., World markets and political instability within less developed countries. Co-operation and Conflict 21 (2), 99-117, 1986.
9. World Bank, Cities, Sea, and Storms: Managing Change in Pacific Island Economies. World Bank, Washington, 2000a.
10. Goldstone, J., Demography, environment, and security, in Diehl, P., Gleditsch, N., (Eds), Environmental Conflict. Westview Press, Boulder, 84-108, 2001.
11. Singh, Jasjit, "India's Strategic and Security Perspective", Vol. XIII, Issue No.5, August 1990, pp. 475-500.
12. Ibid.
13. Dalby, S., Ecopolitical discourse: 'Environmental security' and political geography. Progress in Human Geography, 16 (4), 503-22, 1992.
14. Barnett J., Security and Climate Change, Tyndall Centre for Climate Change Research Tyndall Centre for Climate Change Research Working Paper 7. 2001.
15. Ullah, AKM Ahsan, Climate Change and Climate Refugee in Egypt: An Overview from Policy Perspectives, Accessed through https://fass.ubd.edu.bn/staff/docs/AU/journals/Ullah-2012-climate.pdf2012 on 18 April 2019.
16. The International Institute for Sustainable Development (IISD), Accessed through http://sdg.iisd.org/news/undp-releases-20th-edition-of-human-development-report/on 10 March 2019.
17. Wisner, Ben, *Climate Change and Human Security,* Accessed through http://www.disasterdiplomacy.org/cchswisneretal.pdf 15 April 2007, pp.3-4 on 5 December 2016.
18. Wisner, Ben, *Climate Change and Human Security,* Accessed through http://www.disasterdiplomacy.org/cchswisneretal.pdf 15 April 2007 p. 5 on 6 December2016.
19. Ibid p.6
20. Jauregui, Jason Pareja, What is the effect of COVID-19 on Climate Change?, Accessed through https://www.oneyoungworld.com/blog/what-effect-covid-19-climate-change, accessed on 8 November 2020.
21. Wisner, Ben, *Climate Change and Human Security,* Accessed through http://www.disasterdiplomacy.org/cchswisneretal.pdf 15 April 2007,p. 6 on 6 December 2016.
22. Wisner, Ben, *Climate Change and Human Security,* Accessed through http://www.disasterdiplomacy.org/cchswisneretal.pdf 15 April 2007, p. 8 on 6 December2016.
23. Ibid
24. Levine Simon, Katie Peters and Lilianne, Conflict, Climate change and politics, Humanitarian Policy Group Working Paper. Accessed through https://www.odi.org/sites/odi.org.uk/files/odi-assets/publications-opinion-files/8825.pdf on 14 June 2019.
25. Ibid, p. 9.
26. Gleick, Peter and Iceland Charles, Water, Security and Conflict. Accessed through https://www.wri.org/publication/water-security-and-conflict on 22 August 2019.
27. Water, Security and Conflict. Accessed through https://pacinst.org/wp-content/

uploads/2018/08/Water-Security-and-Conflict.pdf on 11 November 2020.

28. Planetary Security Initiative, Water, Climate and Conflict: security risks on the increase? Accessed through https://www.clingendael.org/sites/default/files/pdfs/Briefing_Note_PSI_Water_climate_and_conflict.pdf on 13 December 2018.
29. World Resources Institute, RELEASE: Significant Risk of Water-Related Conflict in Parts of Iraq, Iran, Mali, Nigeria, India and Pakistan Over Next 12 Months. Accessed through https://www.wri.org/news/2019/12/release-significant-risk-water-related-conflict-parts-iraq-iran-mali-nigeria-india-and on 16 January 2020.
30. Ibid.
31. Quiroz, Andrea Erickson, A shift in perspective: Taking a new view on water security through technology. Accessed through https://blogs.microsoft.com/on-the-issues/2019/09/04/a-shift-in-perspective-taking-a-new-view-on-water-security-through-technology/ on 10 October 2020.
32. Ibid.
33. Ibid.
34. United Nations Climate Change, What is the United Nations Framework Convention on Climate Change? Accessed through https://unfccc.int/process-and-meetings/the-convention/what-is-the-united-nations-framework-convention-on-climate-change, on 25 October 2020.
35. What is the Kyoto Protocol? Accessed through https://unfccc.int/kyoto_protocol, on 28 October 2020.
36. Ibid, p.2.
37. Climate Change, UK government to commit to 2050 target. Accessed through https://www.redmileenergy.co.uk/climate-change-uk-government-to-commit-to-2050-target/ on 8 October 2019.
38. Ibid.
39. Ibid.
40. Ibid, p.1.
41. Ibid.
42. Ibid.
43. Mesjasz, C., Security as an Analytical Concept. Paper presented at the 5th Pan-European conference on International Relations in The Hague. 2004. pp. 911.
44. Ibid.
45. Baldwin, D.A., The concept of security. Review of International Studies, Vol. 23, Issue 01. January 1997. P. 526.
46. Waltz, K., Man, the State and War. New York. Columbia University Press, 1959.
47. Ibid.
48. Ibid.
49. Ibid.
50. MacFarlane, S. Neil, Yuen Foong Khong, Human security and the UN: a critical history. United Nations intellectual history project (illustrated ed.). Indiana University Press, p. 346.
51. Haftendorn, Helga (March 1991). "The Security Puzzle: Theory-Building and Discipline-Building in International Security". International Studies Quarterly (Blackwell Publishing on behalf of The International Studies Association) 35 (1): 3–17.

4

Politics of Climate Change

4.1 What is Politics of Climate Change?

Climate change has moved to the centre stage of public concern in a remarkable way and in a very short space of time. Scientists have expressed serious concerns about global warming for a quarter of a century or more. Environmental groups have struggled for much of that period to get governments and citizens to take the issue seriously. Yet within the past few years, climate change has assumed a very large presence in discussion and debate across the world. Climate change is now a mainstream political issue. However, as yet there is no substantive framework for policy which offers coherence and consistency as to how national governments should cope with the long-term political challenges of climate change.

It is conventional to say of the modern state that it should be an enabling state and this notion certainly applies to the management of climate change policy. A return to planning cannot mean going back to heavy-handed state intervention, with all the problems that it brought in its train. The role of the state (national and local) should be to provide an appropriate regulatory framework that will steer the social and economic forces needed to mobilise action against climate change. I prefer the concept of the "ensuring state" to that of the "enabling state". The idea of the enabling state suggests that the role of the state is confined to stimulating others to action and then letting them get on with it. The ensuring state is an enabling state, but one that is expected or obligated to make sure such processes achieve certain defined outcomes—in the case of climate change the bottom line is meeting set targets for emissions reductions.[1]

One cannot discuss the politics of climate change without mentioning the green movement, which has been a leading influence on environmental politics for many years. It has had a major impact in forcing the issue of climate change into the political agenda, "Going Green" has become more or less synonymous with endeavours to limit climate change.[2] In fact the relationship between the environmental movement and the greening of orthodox politics is a problematic one. Many green groups, for example, have seen themselves as contesting the field mainstream political activity. Some such groups have their origin in conservationism in the idea that we have to protect nature from the ravages of humankind. Yet there can be no question of a "return to nature" as the guiding thread of environmental politics. We live in a world that in many respects is "on the other side of nature" where human intervention into what was the natural world is so profound that there can be no way back. The political and philosophical implications of the retreat of nature are very considerable. They stretch well beyond the area of thinking about climate change but refract back on such thinking in a number of ways. Science is pushing back inner as well as outer nature. Some of these instructions are fiercely resisted as they penetrate the body and human reproductions consider, for example, the heated differences of opinion around embryology today. It is a state that acts in accordance with green values the key problem is to institutionalise them, and in such a way that they accord with other values (such as social justice). Yet we have to decide what green values actually are. Most green parties and organisations are vague on the issue. The Global Greens is an international network of green parties and political movements. It publishes a charter defining what green values are derived from its member groups. However, as laid out in the charter they range all over the place, from ecological wisdom (we acknowledge the wisdom of the indigenous people of world) through to participatory democracy, respect for diversity, sustainability and non-violence. Several of these, such as participatory democracy or non-violence, seem to have nothing at all to do with green values as such. Even sustainability is notoriously difficult to define. A lot more work of clarification is needed in this context.

4.2 Dimensions of Climate Change Politics

- In most of the western countries, political compulsions are how voters perceive policies.
- In India this is hardly an issue. Rather politics of climate change can be divided into four types:
    1. Take India to alternative future: suspect negotiations.
    2. 'Growth-first': suspects climate science.
    3. West is making use of India's stand for not reducing emissions; so, be opportunistic.
    4. Political party dimension: get your work done using climate change agenda.[3]

4.2.1 National Climate Action Plan

- Development of Solar Energy (1000 MW in 2009 to 20,000 in 2020).
- Energy efficiency
- Sustainable habitat
- Water use efficiency
- Sustaining the Himalayan ecosystem
- Building a 'Green' India (afforestation)
- Sustainable agriculture
- Generating strategic knowledge for climate change[4]

4.3 Global-Local

In the area of impacts, climate change has different, although interrelated effects that manifest them at all levels of administration. Considerably, GHG emissions are essentially local in nature, but it is their cumulative concentration level that is a critical factor in causing climate change. While emissions are locally emitted, they are often the result of local through to global processes and policies. At the local level, the effects of climate change are mostly experienced. Changes in climate cause weather patterns to change affecting local agriculture, household practices and vector borne diseases. Policies are being taken from global through to the local level. These policies are not necessarily the result of top-down translation of measures to local level, but are also often the result of external influences from epistemic communities or networks, through which local communities are galvanized into developing possible policy options. Where national governments are reluctant to take action, provincial and local governments have stepped into the vacuum to develop policies.[5]

- Global negotiations assume that local governance will carry out the mandated mission from global negotiations.
- But, in India the 'intermediate' class with huge local political clout is less likely to respect the global negotiation.
- Pressure from below is completely missing.[6]

Climate change has become the major global challenge of this young century. For years, the search for solutions has run up against a sharp North-South divide over the historical emissions of developed countries and the parameters of what is termed, in the climate world, "common but differentiated responsibility" for developing nations. A common appreciation of climate and economic equity between disparate countries and regions remains both critical and challenging for the global climate negotiations process if it is to culminate in a major deal in Paris in 2015, and for implementation beyond that date. The authors believe that the only way to remove this roadblock is to forge an "India exception" in global climate talks; doing so is the only realistic pathway to a global climate deal, and could be a key tool in cementing stronger ties between India and the United States, two critical actors in the evolving international order.[7]

As UN climate talks in Marrakesh enter their final few days, leaders have a lot on their minds. Political supports for a green agenda will wane in America next year. Barack Obama has led international environmental-protection efforts; Donald Trump plans to oppose them. The switch comes just as a new report from the world Meteorological Organisation, a UN body, confirms that year 2016 was the hottest ever recorded. This will mean that 16 of the 17 most sweltering years ever tracked have occurred since the millennium.[8] In the early part of 2016 the climate phenomenon EI Nino, which affects sea temperature in the tropical Pacific and thus rainfall patterns around the world, explained much of the global heat. Some scientists believe the event may also mean that the world's plants have become a less potent carbon sink than in recent years. This is another unwelcome switch. Rising concentrations of greenhouse gases in the atmosphere are warming the world apace. As of this year, levels of carbon dioxide are unlikely to dip below 400 parts per million until technologies to suck the stuff out of the atmosphere are deployed at scale. Millions of years have passed, since the last time the concentration was at this high.[9]

4.4 United Nations Framework Convention on Climate Change (UNFCCC)

The United Nations Framework Convention on Climate Change (UNFCCC or FCCC) is an international environmental treaty that was developed at the United Nations Conference on Environment and Development (UNCED) (informally known as the Earth Summit) in Rio de Janeiro, June 1992. It is basically aimed at stabilizing greenhouse gas concentrations in the atmosphere at a level that would prevent dangerous anthropogenic interference with the climate system commonly believed to be around 2°C above the pre-industrial global average temperature. The treaty as originally framed set no mandatory limits on greenhouse gas emissions for individual nations and contained no enforcement provisions; it is therefore considered legally non-binding. Rather, the treaty included provisions for updates (called "protocols") that would set mandatory emission limits. The principal update is the Kyoto Protocol, which has become much better known than the UNFCCC itself.

Climate change is an entirely natural process that occurs over a wide range of timescales, from a few years to hundreds of millions of years. Recently, there has been a great deal of concern and debate about whether human activity is affecting the temperature cycles of the planet. Global temperatures have risen 0.6°C over the past two centuries alone, a fact that many scientists find alarming.[10]

The countries of the world have taken steps towards creating international agreements to help control the human activities that are believed to affect climate change. Under the umbrella of the United Nations Framework Convention on Climate Change (UNFCCC), the Kyoto Protocol is one such agreement that aims to reduce global emissions of greenhouse gases by the year 2012. There is, however, controversy over the politics behind the agreement and whether it will actually have a measurable, beneficial effect on the climate.[11] The convention puts the onus on developed countries to lead the way. The idea is that, as they are the source of most past and current Greenhouse Gas emissions, industrialized countries are expected to do the most to cut emissions on home ground. They are called Annex I countries and belong to the Organization for Economic Cooperation and Development (OECD). They include 12 countries with economies in transition from Central and Eastern Europe. Industrialized nations agree under the Convention to support for action on climate change.

4.4.1 Objectives of the UNFCCC

According to Article 2, the Convention's ultimate objective is "to achieve, in accordance with the relevant provisions of the Convention, stabilization of greenhouse gas concentrations in the atmosphere at a level that would prevent dangerous anthropogenic [originating in human activity] interference with the climate system". This objective is qualified in that it "should be achieved within a time frame sufficient to allow ecosystems to adapt naturally to climate change, to ensure that food production is not threatened and to enable economic development to proceed in a sustainable manner".

4.4.2 Principles of the UNFCCC

In their actions to achieve the objective of the Convention and to implement its provisions, the Parties shall be guided, inter alia, by the following:

1. The Parties should protect the climate system for the benefit of present and future generations of mankind, on the basis of equity and in accordance with their common but differentiated responsibilities and respective capabilities. Accordingly, the developed country Parties should take the lead in combating climate change and the adverse effects thereof

2. The specific needs and special circumstances of developing country Parties, especially those that are particularly vulnerable to the adverse effects of climate change, and of those Parties, especially developing country Parties, that would have to bear a disproportionate or abnormal burden under the Convention, should be given full consideration.
3. The Parties should take precautionary measures to anticipate, prevent or minimize the causes of climate change and mitigate its adverse effects. Where there are threats of serious or irreversible damage, lack of full scientific certainty should not be used as a reason for postponing such measures, taking into account that policies and measures to deal with climate change should be cost-effective so as to ensure global benefits at the lowest possible cost. To achieve this, such policies and measures should take into account different socio-economic contexts, be comprehensive, cover all relevant sources, sinks and reservoirs of greenhouse gases and adaptation, and comprise all economic sectors. Efforts to address climate change may be carried out cooperatively by interested Parties.
4. The Parties have a right to, and should, promote sustainable development. Policies and measures to protect the climate system against human-induced change should be appropriate for the specific conditions of each Party and should be integrated with national development programmes, taking into account that economic development is essential for adopting measures to address climate change.
5. The Parties should cooperate to promote a supportive and open international economic system that would lead to sustainable economic growth and development in all Parties, particularly developing country Parties, thus enabling them better to address the problems of climate change. Measures taken to combat climate change, including unilateral ones, should not constitute a means of arbitrary or unjustifiable discrimination or a disguised restriction on international trade.

(*Source: United Nations, 1992*)

4.4.3 Responsibilities of the Conference of Parties

The Conference of the Parties, as the supreme body of this Convention, shall keep under regular review the implementation of the Convention and any related legal instruments that the Conference of the Parties may adopt, and shall make, within its mandate, the decisions necessary to promote the effective implementation of the Convention. According to Article 7.2, the COP is responsible for reviewing the implementation of the Convention and any related legal instruments, and has to make the decisions necessary to promote the effective implementation of the Convention. In particular, its role is to:

1. To examine the Parties' commitments in light of the Convention's objective, new scientific findings and experience gained in implementing climate change policies.
2. Promote and facilitate the exchange of information on measures adopted by Parties to address climate change and its effects.
3. Facilitate the coordination of measures adopted by Parties to address climate change and its effects, if requested to do so by two or more Parties.
4. Promote and guide the development and refinement of comparable methodologies for activities related to implementing the Convention, such as preparing inventories of GHG emissions and removals and evaluating the effectiveness of measures to limit emissions and enhance removals.
5. Assess the implementation of the Convention by Parties, the effects of the measures taken by them and the progress made towards achieving the ultimate objective of the Convention.
6. Consider and adopt reports on the implementation of the Convention, and ensure their publication.
7. To make recommendations on any matters necessary for the implementation of the Convention.
8. To seek to mobilize financial resources.
9. To review reports submitted by its SBs and provide guidance to them.
10. To exercise such other functions as are required to achieve the objective of the Convention as well as all other functions assigned to the COP under the Convention. (Source: UNFCCC 2006).

4.5 KYOTO PROTOCOL

The Kyoto Protocol adopted in Kyoto, Japan, on 11 December 1997 is an international agreement that sets binding obligations on industrialized countries to reduce emissions of greenhouse gases and it is linked to the United Nations Framework Convention on Climate Change, which commits its Parties by setting internationally binding emission reduction targets. The Kyoto protocol came into force on 16 February 2005 and it has placed a heavier burden on developed nations under the principle of "common but differentiated responsibilities" by recognizing that the developed countries are principally responsible for the current high levels of GHG emissions in the atmosphere as a result of more than 150 years of industrial activity. Developing countries do not have binding targets under the Kyoto Protocol, but are still committed under the treaty to reduce their emissions. During the first commitment period, 37 industrialized countries and the European Community committed to reduce GHG emissions to an average of five percent against 1990 levels. During the second commitment period, Parties committed to reduce GHG emissions by at least 18 percent below 1990 levels in the eight-year period from 2013 to 2020; however, the composition of Parties in the second commitment period is different from the first. The detailed rules for the implementation of the Protocol were adopted at COP 7 in Marrakesh, Morocco, in 2001, and are referred to as the "Marrakesh Accords." Its first commitment period started in 2008 and ended in 2012.[12] The protocol was developed under the UNFCCC—the United Nations Framework Convention on Climate Change. Participating countries that have ratified the Kyoto Protocol have committed to cut emissions of not only carbon dioxide, but of also other greenhouse gases, being: Methane (CH_4), Nitrous oxide (N_2O), Hydro fluorocarbons (HFCs), Per fluorocarbons (PFCs), Sulphur hexafluoride (SF_6). If participant countries continue with emissions above the targets, then they are required to engage in emissions trading; i.e. buying "credits" from other participant countries that are able to exceed their reduction targets in order to offset. The goals of Kyoto were to see participants collectively reducing emissions of greenhouse gases by 5.2 per cent below the emission levels of 1990 by 2012.[13]

The Kyoto Protocol applies to industrialized nations only.

Developing countries, including India and China, were not required to commit to reductions because their per-capita greenhouse gas emissions are much lower than those of developed nations. The plan is that poorer countries will be brought more actively into future climate change agreements as cleaner technologies develop and become less expensive. Parties to the Kyoto Protocol adopted an amendment to the Kyoto Protocol by decision 1/CMP.8 in accordance with Articles 20 and 21 of the Kyoto Protocol, at the eighth session of the Conference of the Parties serving as the meeting of the Parties to the Kyoto Protocol held in Doha, Qatar, in December 2012.[14] The major distinction between the Kyoto Protocol and the United Nations Framework Convention on Climate Change is that while the Convention encouraged industrialised countries to stabilize greenhouse gas emissions, the Protocol commits them to do so.

Recognizing that developed countries are principally responsible for the current high levels of emissions in the atmosphere as a result of more than 150 years of industrial activity, the Protocol places a heavier burden on developed nations under the principle of "common but differentiated responsibilities."[15] Under the Treaty, countries must meet their targets primarily through national measures. However, the Kyoto Protocol offers them an additional means of meeting their targets by way of three market-based mechanisms:

- *The Emissions Trading:* It is known as the carbon market it is envisaged as the key tool to reduce GHG emissions at global level.
- *The Clean Development Mechanism (CDM)*: It is a project-based mechanism that feeds the carbon market and involves investment in sustainable development projects that reduce emissions in developing countries.
- *The Joint Implementation (JI):* It is a project-based mechanism that feeds the carbon market and helps stimulate green investment by encouraging industrialized countries to carry out joint implementation projects with other developed countries.

According to UNFCCC, the Kyoto mechanism help to stimulate sustainable development through technology transfer and investments help countries with Kyoto commitments to meet their targets by reducing emissions or removing carbon from the atmosphere in other countries in a cost-effective way and encourage the private sector and developing countries to contribute to emission reduction efforts. These Kyoto mechanisms enhance the flexibility of Annex I Parties to meet their emission reduction or limitation commitments, by allowing these Parties to take advantage of lower-cost emission reductions outside their territories. IPCC[16] in its fourth assessment report performed the assessment of international agreements on climate change and highlighted various key issues involved in it as shown in Table 2.

The UNFCCC and Kyoto Protocol agreements currently govern the international response to climate change and clearly put the obligation on industrialized countries to take the lead in combating climate change and its adverse effects. Under the current negotiations developing countries are coming under increasing pressure to commit to binding targets for their own GHG emissions. Countries from around the world are working together to share technologies, experience, resources and talent to lower net greenhouse gas emissions and reduce the threat of global climate change. As countries continue to grow and develop, international co-operation becomes increasingly important as the global community searches for ways to meet the climate challenge efficiently and effectively. The key to successful co-operation is finding activities that will help all countries achieve their economic, environmental and developmental goals in a climate friendly manner. Challenges and emergence of global environmental problems impose an unprecedented requirement for international co-operation. Obviously, a global dimension of climate change problem requires equally global solutions. The chosen way the international community has pursued hovers on how and possibly jointly to respond to the unexpected future common threats through negotiation and action plans to combat global warming.

As scientific data on global warming have gradually accumulated, research institutions and political leaders around the world have recognized that the effective measures to tackle global climate change to prevent such warming trends must be developed through international co-operation. The climate discourse is changing to reflect

Table 2: Assessment of International Agreements on Climate Change

Approach	*Environmental effectiveness*	*Cost effectiveness*	*Meets distributional considerations*	*Institutional feasibility*
National emission targets and international emission trading (including offsets)	Depends on participation, and compliance	Decreases with limited participation and reduced gas and sector coverage	Depends on initial allocation	Depends on capacity to prepare inventories and compliance. Defections weaken regime stability
Sectoral agreements	Not all sectors amenable to such agreements, limiting overall effectiveness. Effectiveness depends on whether agreement is binding or non-binding	Lack of trading across sectors increases overall costs, although may be cost-effective within individual sectors. Competitive concerns reduced within each sector	Depends on participation. Within-sector competitiveness concerns alleviated if treated equally at global level	Requires many separate decisions and technical capacity. Each sector may require cross-country institutions to manage agreements
Coordinated policies and measures	Individual measures can be effective; emission levels may be uncertain; success will be a function of compliance	Depends on policy design	Extent of coordination could limit national flexibility; but may increase equity	Depends on number of countries (easier among smaller groups of countries than at the global level)
Cooperation on Technology R&D	Depends on funding, when technologies are developed and policies for diffusion	Varies with degree of R&D risk Cooperation reduces individual national risk	Intellectual property concerns may negate the benefits of cooperation	Requires many separate decisions. Depends on research capacity and long-term funding

(Contd.)

Table 2 (*Contd.*)

Approach	*Environmental effectiveness*	*Cost effectiveness*	*Meets distributional considerations*	*Institutional feasibility*
Development-oriented actions	Depends on national policies and design to create synergies	Depends on the extent of synergies with other development objectives	Depends on distributional effects of development policies	Depends on priority given to sustainable development in national policies and goals of national institutions
Financial mechanisms	Depends on funding	Depends on country and project type	Depends on project and country selection criteria	Depends on national institutions
Capacity building	Varies over time and depends on critical mass	Depends on programme design	Depends on selection of recipient group	Depends on country and institutional frameworks

Source: IPCC 2007.

a) The table examines each approach based on its capacity to meet its internal goals not in relation to achieving a global environmental goal. If such targets are to be achieved, a combination of instruments needs to be adopted. Not all approaches have equivalent evaluation in the literature; evidence for individual elements of the matrix varies.

b) Research, Development and Demonstration

more grave assessments of the problem in recent years. Some climatologists think that shift is concordant with science community understanding of the nature of the problem ('alarming'), and some think that the shift is rhetorical and inconsistent with the science ('alarmist'). We need to develop as good a sense of the threat as we can get in the limited time available and choose a discourse that sets the appropriate course. Measures to combat global warming must be based upon scientific analysis and knowledge of greenhouse gas emissions and reduction potentials to deepen mutual understanding. It is necessary for all countries to further dialogue and interaction to strengthen their combined efforts against climate change. During Trump's presidentship, the issue of climate change remained under the shadow and became sluggish. The outgoing US president; Donald Trump constantly denied climate change and its urgent demand of addressing. The US participation in the global attempts towards mitigating Climate change headed by the UN, gradually shortened. Finally, the US officially withdrew from the Paris Climate Agreement on November 4, 2015. The agreement aims at confining the global warming at the level of 2°C that is above pre-industrial levels.

4.5.1 Davos 2020

Going by the media coverage, Davos 2020 was the Trump and Greta show.

The Swedish teenager Thunberg, for a second year running, told business and governments to make drastic and urgent cuts in greenhouse gas emissions to avoid the climate disaster that scientists say we are rapidly heading towards. The US president did not mention climate change and said we should ignore the "perennial prophets of doom" and "embrace the possibilities of tomorrow".

4.5.2 COP24

At the UN climate summit 2018 (COP24) in Katowice, Poland, the Global Carbon Project issued a report which estimates a rise in CO_2 emissions by 2.7 percent in 2018 after a 1.6 percent increase in 2017. The key massages were focused on three issues: technology, human and nature. In October 2018, the Intergovernmental Panel on Climate Change released a report that global carbon emissions would need to fall 50 percent by 2030 and reach net zero by 2050. In November 2018, the UN came with a different report on "emission gap" showing the difference between the actions countries are taking to reach the Paris climate targets. Mainly it was suggesting that 'current Paris commitments would need to be tripled to reach a 2°C target and increased fivefold to reach the 1.5°C target'.[17] Climate change and global warming has been alarming the world. Climate change has the potential to be the greatest security threat in this century. It is considered a multiplier of a range of security threats from conflict exacerbation to migration, water, food and health security. This report is designed to equip readers with both conceptual and practical tools for understanding climate change and its impact on the global, regional and local security environment and devising possible solutions to address this multifaceted challenge.

There is growing momentum globally to view climate change as a security issue. Climate change is recognized as a threat to the international security and cooperation. In theoretical terms, it is being securitized and included in the traditional security studies. Climate change and its accompanying environmental degradation are perceived to be a threat that can have incalculable consequence on the international community. In recent years, the security implications of the climate change are being addressed from national, regional and multilateral level. There are at least seven ways that climate change is likely to affect security in its narrow and wider meanings: food-livelihood-climate-conflict, health-climate-livelihood-conflict-security, mitigation-adaptation-inequity-conflict-security, food-fuel conflict and human security, disaster-livelihood-governance-conflict-security, water-conflict-security, sea level-displacement-security.

Climatic variability, including changes in the frequency of extreme events (like droughts, floods and storms), has always had a large impact on humans. The environment is changing rapidly. Increase in world population, accompanied by industrialization and other human activities are altering the atmosphere, ocean, land, ice cover, ecosystems, and the distribution of species over the planet. Understanding these and other global changes, including climate change, is critical to nation's health and economic vitality. Scientific research is critical in gaining this understanding.

Over the last two decades or so, there has been growing concern from the international community over possible security implications of the global climate change. The United Nations (UN), European Union (EU), United States (US), African Union and many other regions of the world have recently begun to ask how the security dimensions of climate change will affect-access to shared resources, national security, borders integrity, trade routes and international political alliances and economic relations. By maintaining international peace and security, the UN makes all its work more effective, because by keeping the peace, the organization can focus on solving global issues, instead of resolving conflicts.

Climate change is a far greater threat to the world than international terrorism (David King, the UK Government's chief scientific adviser, 2004). It may spark conflict between nations. John Reid foresaw violence and political conflict more likely to occur in the next 20 to 30 years as climate change turned land into desert, melted ice fields and poisoned water supplies. He listed climate change alongside the major threats in future decades, including terrorism, demographic changes, and global energy demand (John Reid MP, the then British Secretary of State for Defence and now Home Secretary). John Ashton, the UK Foreign Secretary's Special Representative for Climate Change, said at a conference on "Climate Change: The Global Security Impact", at the Royal United Services Institute on 24 January 2007: "There is every reason to believe that as the 21st century unfolds, the security story will be bound together with climate change. Climate change is a security issue because if we don't deal with it, people will die and states will fail".[18]

Trump as a businessman, vocally, has stated internationally that "climate change" does not exist. The article entitled 'On Bullshit', by philosopher Harry Frankfurt, published in 1986, used the term bullshit to define 'the presence of insincerity, laziness towards details and half-heartedness an indifference to truth'. Of late, there were a number of reports proclaiming that in comparison to non-committed firms, the one who committed to environmental, social and governance (ESG) financially performing better. Under such condition, Larry Fink, CEO of Black Rock, denied the directors from investing to the firms not producing their climate goals report, compelled the private sectors to do so.

Concurrently, the need for national climate and environmental related obligations became robust. Eventually, in 2015, 195 countries joined together and committed for climate action. To bring down the global temperatures, each country agreed to reduce their greenhouse gas emissions voluntarily, came to known as Paris Agreement. Despite the commitment by 195 countries to reduce greenhouse gas emissions, accomplishing climate goal for the world will be insufficient since the countries committed to Paris agreement, are also the major subsidisers of fossil fuels. Therefore, it needs to build governance mechanism which includes independent verifiers who report and verify these environmental and social gains separately to a regulatory authority. The US has failed to fulfil its promises to reduce greenhouse emissions under his rule and also exited from the crucial Paris agreement. As we have discussed in the report earlier, the Paris agreement is important step for climate change issue, because 2020 is considered as the tipping point for the world in terms of global warming. If countries fail to take action by 2020, it might be too late to save the world from a climate crisis. Hence, when the US withdrew from the Paris agreement as Trump did not favour it, things have gone from bad to worse on the climate issue. Trump has not only turned itself away from the Paris Agreement commitments but several other decisions. Global warming is real, is happening now and is the result of human activities. Barack Obama and Joe Biden believe we have a moral, environmental, economic and security imperative to tackle climate change in a serious, sustainable manner. Unlike Donald Trump, Barack Obama and Joe Biden, support implementation of a market-based cap-and-trade system to reduce carbon emissions by the amount scientists say is necessary: 80 percent below 1990 levels by 2050. Under the Biden administration, he intends to start reducing emissions immediately by establishing strong annual reduction targets as well as implement a mandate of reducing emissions to 1990 levels by 2020.

NOTES

1. Giddens, Anthony, The politics of climate change National responses to the challenge of global warming, Accessed through http://www.policy-network.net/uploadedFiles/Publications/Publications/The_politics_of_climate_change_Anthony_Giddens.pdf on 15 December 2016 p9.
2. Giddens, Anthony, The Politics of Climate change, Polity Press Reprinted, 2009.
3. Pellissery Sony, "India and the Politics of Climate Change", Institute of Rural Management, Anand Gujarat, India, Accessed through http://www.iddri.org/Evenements/Conferences/100323_presentation_India%20and%20the%20Politics%20of%20Climate%20Change.pdf on 15 November 2016.
4. Ibid
5. Gupta, Joyeeta, Kim van der Leeuw and Hans de Moel, Climate Change: a 'Glocal' Problem Requiring 'Glocal' Action. Accessed through, https://www.tandfonline.com/doi/pdf/10.1080/15693430701742677 on 3 July 2019.
6. Ibid.
7. Jones Bruce and Saran Samir, "An 'India Exception' and India-U.S. Partnership on Climate Change," *Planet Policy*, Monday 12 January 2015, Accessed through https://www.brookings.edu/blog/planetpolicy/2015/01/12/an%20-india-exception-and-u-s-partnership-on-climate-change/ on 18 November 2016.
8. The state of the climate in 2016, Accessed onwww.economist.com/blogs/graphic detail/2016/11/global-warming on 20 November 2016.
9. Ibid.
10. Climate Change and the Kyoto Protocol—The Kyoto Protocol, December 11, 1997, Accessed through http://climatechange.sea.ca/kyoto_protocol.html on 30 December2016.
11. Ibid.
12. UNFCCC. Accessed through: http://unfccc.int/kyoto_protocol/items/2830.php on 1December 2016.
13. Ibid.
14. United Nations Framework Convention On Climate Change, 5 October 2016, Accessed through http://unfccc.int/kyoto_protocol/items/2830.php on 17 December2016.
15. The Kyoto Protocol: Environmental Protection Agency, Ireland. Accessed through http://www.epa.ie/climate/thekyotoprotocol/ on 27 December2016.
16. IPCC (2007). Climate Change 2007: Mitigation of Climate Change: Contribution of Working Group III to the Fourth Assessment Report of the Intergovernmental Panel on Climate Change. Cambridge, UK and New York: Cambridge University Press.
17. Chelsea Harvey, CO_2 emissions reached all-time high in 2018, Scientific American, 6 December 2018. Accessed through https://www.scientificamerican.com/article/co2-emissions-reached-an-all-time-high-in-2018/. Accessed on 12 May 2019.
18. Wisner, Ben, Climate Change and Human Security, Accessed through http://www.disasterdiplomacy.org/cchswisneretal.pdf on 5 December2016.

5

Climate Change and Pandemics

5.1 Indian Perspectives on National Security

The concept of national security is much wider than that of defence, and the approach to national security requires comprehensive views of various political social economic, technological, and strategic aspects.[1] There is a widespread impression of the term National Security. It is self-evident and it involves safeguarding our territorial integrity and sovereignty. Sovereignty and abridgement are admittedly sophisticated concepts, while defence of territorial integrity is simple enough. Those familiars with Western literature will define the concept of national security as the ability of a nation to protect its internal values from external threats. Walter Lippmann the noted columnist once defined it in following terms. A nation has security when it does not have to sacrifice its legitimate interests to avoid war and is able if challenged, to maintain them by war, while these definitions adequately cover the concept as applied to stabilize nation state of the industrialized west. These are in adequate to encompass the entire area of the problem of national security faced by such a heterogeneous and developing nation state like India.[2] Conventional wisdom in most parts of the world has tended to threat national security as synonymous with national defence. The reality, which is being increasingly appreciated, is that the former covers a much broader spectrum of challenges, threats and responses as compared to the latter. Which being a sub-set of national security in its comprehensive framework, relates to military security essentially from external threats. At the same time it has to be noted that capabilities also constitute the ultimate instrument in application of force when and where required in the pursuit of national security.[3]

In India the broader concept of National Security may be seen to have it roots in the struggle for Independence itself, and well-articulated and operationalized in the early years of India's history as an Independent Nation state. Nehru's concept of an 'area of peace' pursued the policy of non-alignment in a global system. The basic approach also emphasized that India's security issues cannot be isolated from the geo-strategic environment and geopolitical realities of the international system. In this context, therefore, it would be useful to attempt a definition of national security so as to formulate the framework of national strategy to achieve it. According to Morton Berkowitz and P.G. Bock, national security can be most fruitfully defined as the ability of a nation to protect its internal values from external threats. However, this definition falls short of a comprehensive approach to national security. Views on exact definitions will no doubt differ, but we may adopt a broad concept of national security accordingly changes took place in international security environment, i.e. the preservation of the core values critical to the nation state from external and internal threats. According to this definition to identify the core values at the same time it is necessary to note that although external and internal security issues are substantially different in nature. Many issues affecting internal security may stem external, political, social, economic or military pressures. Similarly domestic insecurity and perhaps what are even more relevant, perceptions of domestic vulnerabilities may create incentive for external pressures.[4] India has always approached the subject of security in its larger framework, beyond that implicit in defence and military forces. The concept of security has involved the preservation and perpetuation of the core value, i.e. democratic political set-up, secular state, socialism, nature of the state, attainment of egalitarian society, maintenance of internal peace and security and economic development and progress. These values had shaped the Indian civilization and they provide the foundation on which modern Indian can be built taking into account the historical socio-economic condition in the country. Given the pluralistic society socio-economic inequities, regional disparities at the time India became independent these core values stand enshrined in the Indian constitution since 1950.

The traditional security thinking based on military capability to cope with external military threat, therefore, needs to be re-oriented to meet the challenges and threats. The 19th century concept of a nation in arms moved on to that of a nation at war in the first half of the 20th century as witnessed during the two world wars. Additionally due to coercive strategy in the nature of external threats requires new approaches to formulation of national security doctrines and strategy for a nation under threat of war. These essentially demand greater interactive foreign, domestic and defence policies for a comprehensive national security strategy. These fundamental issues need to be taken into account in any discussion of the national security.

The core values of the Indian nation state derive their strength from its culture and civilization. In the conceptual framework for India's national security the three main core values stands for: Democracy, Secularism and Egalitarian Society, erosion of any or all of these values in any form for whatever reason, erodes the legitimacy of the Indian nation-state and hence poses threats to its security. It was for this reason that these values were so carefully enshrined in the constitution of India.[5] The term security of a state generally means the preservation of territorial integrity, political independence, sovereignty and the economic autonomy of the people. As a concept, it has long been contested, and has remained underdeveloped. However till the end of the Cold War, the concept was associated with the threats or use of force, force and associated concepts or strategies like war, limited war, deterrence, alliances, and arms control disarmament and crisis management.

But after the demise of Soviet Union, the end of cold war, and the diminishing chances of war or nuclear war, emphasis has been on rethinking its content and defining it more broadly and present day in the age of globalization security problems, especially for developing countries like India are result of terrorism, trans border solidarity, malnutrition, illiteracy, poverty, social disparities, lower gross domestic product, imbalance market economy, communalism, violations of human rights, etc., which were traditionally kept out of national security discourse. The fundamental challenge before us today is how to include the domestic challenges before a state for broadening the agenda of national security. Internal challenges to the state have shown its pervasiveness in different parts of the world. Traditionally it is argued that threats emanate domestically from different groups, competing for power in a political system. Where consensus is often totally absent for example, Yugoslavia, Somalia, Bosnia, Afghanistan, Congo and Sri Lanka where particularly the ethno religious and political rivalries have significant ramification and at several levels they have fused with international crisis. After the end of the Cold War experiences at global level would suggest that inter challenges to security are rooted in problems of political identity, socio-economic grievances, lack of legitimacy of state and individuals, and the most important reason is the non-stability of political and economic institutions.[6] The idea of security has undergone changes keeping pace with the changing international security environment. From the narrow definition of security of territory from external aggression to global implies the threat of a nuclear holocaust, security today implies among other things, protection from threats of disease, hunger unemployment, crime, social conflicts, political repression and environmental hazards. In recent years the UN's concern for social issues and development has also be receiving considerable attention. However, UN interest in social progress is not new, it derives directly from its character, "To promote higher standards of living full employment and conditions of economic and social progress and development."[7] We emphasize more for our national security priorities, hardly considering internal security consolidation as an important matter, on the threats from Pakistan and china. Such a restricted view focused on limited issue, i.e. foreign policy, military capabilities, defence planning, and defence expenditure the articulation of social security needs have been left to the socially concerned, whose voice is not heard in the security arena. If asked, a layman would probably explain national security as the protection of territorial integrity, which for that purpose would require the adequate building up of the armed forces along with the appropriate acquisition of military hardware. This view, however, is simplistic and does not address and complexities forced by nation-state today. After all, in addition to securing territorial integrity, the people of a country would neither like its political or constitutional system to be challenged nor want its economic, technological or scientific progress and interest to be in any manner of jeopardy. A nation would like to safeguard and promote its core values, which could be termed the nation's way of life, such as democracy, federalism or human rights.

The term national security would, therefore have to encompass all these factors. If that were so, then it cannot be assumed that challenges to a country's national security emanate only from the external environment. A useful definition, which is relevant to present times, state "security is not military hardware though it may include it, security is not military force, security is development and without development there is no security". Another analyst suggests that national security does not merely mean safeguarding territorial integrity. It also means ensuring that the country is industrialized rapidly and has a cohesive, egalitarian and technological society. Anything that comes in the way of this development internally or externally is a threat to national security. At this stage it has to be noted that challenges to national security have to be combated by one or more or a combination of the instrumentalists available to the state, i.e. domestic capabilities, diplomatic avenues, military strength. Similarly these tools of the state are used not only in a defensive posture, but also to further national interests. These three state capabilities draw their ability or strength or the lack of it from the elements of national power, which is a mix of strategic, military, economic and political strength and weakness. It is determined in part by military forces but even more by the size and location of territory, the nature of its frontiers, population, economic structure, natural resources, technological developments, social cohesiveness, institutions, stability of its political process and finally the national character and spirit. The foregoing, in a sense has already generated the need for deeper analysis of the theoretical and empirical factors which determine security for nation within itself, in the external environment and in relation to all other nations. Thus, it may suggest that basically there exists in the very core of national security three levels or three separate constituents. These are in ascending order the individual, the state and the international order or system in it's entirely. The importance of these three levels of national security lies in their intrinsic security values and equally in the obvious connections between individual or personal security that of the state and the security of the international system.

Mallik A demonstrated a technology perspective of the security concerns and challenges for India, in the complex international environment of simultaneous competition and cooperation.[8] India's sustained techno-economic progress, combined with its performance in the areas of strategic technology and information-communication technology, have catapulted the country to the status of a global technology player and a strong candidate for inclusion among the top world powers, in the near future. The role of technology in India's transformation has been significant. Sustaining this momentum and achieving numerous national objectives in an atmosphere of changing security dynamics, will be the real challenge.[9]

5.2 Health and Environment

Health problems related to the environment continue to be a major source of concern all over the world. Society needs to develop measures that will eliminate or considerably reduce hazardous factors from the environment that can result in health risk to humans. Catastrophic losses in biodiversity, reckless destruction of wild-land and warming temperatures have allowed disease to explode. Ignoring the connection between climate change and pandemics would be "dangerous delusion,"[10] one scientist said. There is scientific consensus worldwide through the Intergovernmental Panel on Climate Change, and National Academy of Sciences of USA, and other scientific professional associations, that the human modification of the global climate is occurring and that the climate is warming. Human activity is changing the climate by changing the chemical balance in the air. Certain gases in the atmosphere, called 'greenhouse gases,' absorb outgoing long wave radiation and re-emit back to the surface heat that would otherwise escape to space. The atmospheric concentrations of several greenhouse gases emitted by human activities, including carbon dioxide and methane, are increasing dramatically. Because these gases are causing the atmosphere to recycle heat and hold in more warmth, the average temperature at the surface of the planet is going up.

The rate of increase has accelerated in the last 50 years, and significant further increases are expected in the next century and beyond. The global average surface temperature has increased over the twentieth century by about 0.6°C, and is projected to rise by an additional 1.4–5.8 degree Cover the twenty-first century, depending on future emissions of greenhouse gases. Over the past few decades, the number of emerging infectious diseases that spread to people—especially coronaviruses and other respiratory illnesses believed to have come from bats and birds has skyrocketed. A new emerging disease surfaces five times a year. One study estimates that more than 3,200 strains of coronaviruses already exist among bats, awaiting an opportunity to jump to people. The diseases may have always been there, buried deep in wild and remote places out of reach of people. But until now, the planet's natural defence systems were better at fighting them off. Today, climate warming is demolishing those defence systems, driving a catastrophic loss in biodiversity that, when coupled with reckless deforestation and aggressive conversion of wildland for economic development, pushes farms and people closer to the wild and opens the gates for the spread of disease. There are three ways climate influences emerging diseases. Roughly 60 per cent of new pathogens come from animals including those pressured by diversity loss and roughly one-third of those can be directly attributed to changes in human land use, meaning deforestation, the introduction of farming, development or resource extraction in otherwise natural settings. Vector-borne diseases those carried by insects like mosquitoes and ticks and transferred in the blood of infected people are also on the rise as warming weather and erratic precipitation vastly expand the geographic regions vulnerable to contagion. Climate is even bringing old viruses back from the dead, thawing zombie contagions like the anthrax released from a frozen reindeer in 2016, which can come down from the arctic and haunt us from the past.

Thus, the COVID-19 pandemic, even as it unfolds in the form of an urgent crisis, is offering a larger lesson. It is demonstrating in real time the enormous and undeniable power that nature has over civilization and even over its politics. That alone may make the pandemic prologue for more far-reaching and disruptive changes to come. But it also makes clear that climate policy today is indivisible from efforts to prevent new infectious outbreaks or the notion that climate and health and environmental policy might not be related is "a dangerous delusion."[11]

A heating planet might increase the frequency of infectious diseases.

Rising global temperatures, coupled with the increasing frequency of extreme weather events, are predicted to cause changes in the seasonality, geography and intensity of infectious diseases. Floods can enhance the spread of infectious agents like insects, bacteria, and viruses. Increasing temperatures and humidity affect the development, survival and spread of not only pathogens but also their hosts (often animals). Mosquitoes and other insects that are carriers of diseases like malaria, dengue fever and the West Nile virus will move to areas of the planet that are colder today, as they warm up. The West Nile virus, for example, first appeared in the northern hemisphere in New York, after a prolonged period of high temperatures followed by heavy rainfall in 1999. As Stanford University professor of biology Erin Mordecai has pointed out, "wealthy, developed countries such as the US are not immune."

5.3 Climate Change and Public Health

The link between climate change and public health has received increasing attention on the policy front at the international and national level. The 2016 Global Risk Report published by the World Economic Forum identifies the link between climate change and the spreading of infectious diseases, which are two of the top ten high impacts, high frequency risks highlighted in the report. Also, in early April 2016, the US Global Change Research Program (USGCRP) released a report that examines the various ways through which climate change will affect overall public health. Changes in climatic conditions will act as threat multipliers to a wide range of non-communicable and infectious diseases, from temperature-related deaths, asthma and allergy conditions, water-related diseases such as cholera and meningitis, to mental health and stress-related illnesses. There are multiple channels through which climate change will directly and indirectly affect the frequency and intensity of pandemic outbreaks. Due to the highly complicated and multidimensional nature of these links, applying analytical and holistic frameworks improves understanding of these various interconnections. Applying the One Health concept, which recognizes the connections between environmental, animal and human health, helps identify the mechanisms through which changes in the environment can affect the spreading of infectious diseases. The Haddon Matrix facilitates analysis of factors leading up to the event, as well as those that affect the severity of an outbreak. Modified versions of the two frameworks will be applied in this report to identify the first, second, and third order effects that climate change will have on pandemic risk.

The effect on human health can be measured in terms of vulnerability, which is defined by USGCRP as "the tendency to be adversely affected by climate-related health effects."[12] Three key elements of vulnerability to infectious diseases are: Exposure to vectors and disease carrying agents, sensitivity to climate change and its health effects, and adaptive capacity of communities. Climate change affects vector ecology and the vulnerability of human populations before an outbreak, and exacerbates the intensity during an outbreak. Prior to an outbreak, changes in temperature, precipitation patterns and pH levels will affect the quantity and quality of ecosystem services such as provisioning services (food supply, water availability, etc.) and supporting services (soil formation, nutrient recycling, etc). This will in turn affect migratory patterns and habitat of certain animals and insects. Moreover, changes in precipitation patterns, atmospheric temperature, and seasonality will influence animal and insect populations and their survivability. Climate change can also affect sensitivity if it negatively affects overall health and nutrition conditions of vulnerable populations in developing countries where food and water security are at risk. These fundamental conditions are related to how susceptible a population or individual may be to the spreading of infectious diseases.

A pandemic outbreak could occur after a natural disaster and/or a sudden collapse in regulating services, such as pest control or water purification. The 1976–77 droughts in California, for example, reduced the population of rodent predators. The subsequent rainfall led to an increase in plant growth, which in turn increased rodent populations as well as the transmission of Hantavirus Pulmonary Syndrome (HPS) a fatal disease with a 38 per cent mortality rate. If a pandemic outbreak occurred during periods when there is increased activity of vectors in regions where human-human interactions and transmission levels are high, such as urban areas, then the impact of an outbreak will be even more severe. Comprehending the enormity of climate change is about as mind-bending as understanding the ultimate effect novel coronavirus

SARS-CoV-2 (the virus that causes Covid-19) will have on the world. COVID-19 could well be nature's warning against climate change because microbes are the first to ring an alarm and this could just be the first wave of the pandemic with bigger outbreaks likely to follow.

5.4 COVID-19 AND CLIMATE CHANGE

History shows that pandemics such as COVID-19 typically last for years, and come in recurring waves. The Spanish Flu of 1918, one of deadliest in history, was far more fatal in the second wave. It affected one-third of the global population and almost certainly altered the course of the ongoing First World War. In some ways, the dire lockdowns undertaken to stop COVID-19 have fast-forwarded us into an unlikely future—one with almost impossibly bold climate action taken all at once, no matter the cost. Just months ago it would have been thought impossible to close polluting factories virtually overnight and slash emissions from travel by keeping billions at home. Now we know that clear skies and silent streets can come about with shocking speed. The pandemic is a cataclysmic event so big and disruptive that it can be measured in the planetary metrics of climate change. As many as 2.6 billion metric tons of carbon dioxide emissions, about 8 per cent of the estimated total for the year, will never be emitted into the atmosphere, according to estimates by the International Energy Agency. The lockdowns have brought the tourism and travel sectors to a halt. With companies conducting a giant "meetings-that-could-have-been-emails" experiment, analysts and experts wonder whether the travel sector will ever fully recover. Before the pandemic, aviation accounted for more than 2 per cent of global emissions. Detecting the swift drop in pollution has underscored the extent to which it made us vulnerable to the virus in the first place. The burning of fossil fuel in vehicles, factories and power plants creates soot and other particles so small that they end up in almost every organ in the body, including the lungs. COVID-19 is a respiratory illness that can cause pneumonia and result in severe, long-lasting lung damage. People infected with the novel coronavirus are more likely to die if they live in regions with high levels of pollution, according to two different studies. These early findings resonate with pre-pandemic research linking pollution to a long list of diseases. Seven global cities—

Delhi, Sao Paulo and New York among them—experienced 25 per cent to 60 per cent reduction in fine particulate matter known as PM2.5 during the lockdown, according to IQ Air.

Los Angeles experienced its longest stretch of clean air on record, finally meeting the World Health Organization's air quality guidelines. What makes the breathtaking, planetary-scale transformations wrought by the pandemic so unsettling is that none of it registers in the biggest picture. The photographs of clear skies, the charts of falling emissions, the change in daily behaviour by billions of people—none of this will slow the dangerous pace of global warming. The level of CO_2 in the atmosphere, which drives our warming temperatures, has been rising steadily by an average of almost 2.5 parts per million a year since 2010. A significant cut in new emissions from 2020 won't improve this all-important metric, now at 414 ppm. Even a 10% drop in emissions from this year would still translate to an increase of 2 ppm in the CO_2 count. Because of the inertia in the climate system, even if we were to significantly reduce or stop our emissions today, you would still see the increase in temperatures expected for the next 20 years almost unaffected. In reality it is very likely that the total concentration of CO_2 in the atmosphere will continue going up in the future.[13]

NOTES

1. Subrahmanyam K., "Perspective in defence planning", Abhinav Publications, New Delhi, 1972, p.7.
2. Ibid, p. 1.
3. Singh Jasjit, "India's Strategic and Security Perspective", Vol.XIII, Issue No. 5August 1990, pp. 475-500.
4. Ibid.
5. Ibid.
6. Mishra Rajesh K., "Rethinking India's National Security: A Political Approach", Journal of Peace Studies, Vol. 5, Issue No.2, April 1998, pp. 26-32.
7. Rao Meena and Vijayalakshmi, "Challenges of Peace and Security in the 21st Century", Third Concept, Vol. 131, Issue No. 11, January 1998, p. 7.
8. Mallik A., National security challenges and competition for India: Defence and space R&D in a strategic context. 2008. Technology in Society, Volume 30, Issues 3–4, p.362–70.
9. Ibid.
10. Lustgarten, Abrahm, How Climate Change Is Contributing to Skyrocketing Rates of Infectious Disease. Accessed through https://mahb.stanford.edu/library-item/

how-climate-change-is-contributing-to-skyrocketing-rates-of-infectious-disease/ on 18 November 2020.

11. Ibid.
12. Estrada, Rachel et al, Pandemics in a Changing Climate Evolving Risk and the Global Response Accessed through https://www.swissre.com/dam/jcr:552d59b2-76c6-4626-a91a-75b0ed58927e/Pandemics_in_a_changing_climate_Full_ report_FINAL.pdf on 20 November 2020.
13. Lombrana, Millan, Laura and Hayley Warren, A Pandemic That Cleared Skies and Halted Cities Isn't Slowing Global Warming. Accessed through https://www.bloomberg.com/graphics/2020-how-coronavirus-impacts-climate-change/ on 22 November 2020.

6

India-US Strategic Partnership and Perceptions on Climate Change

6.1 India–US Relationship

Humans are altering the Earth's climate. The problem is global. The greenhouse emissions of one country impact all countries. The solution to such a global problem can only come through international cooperation. Cooperative efforts to solve the problem are under way. However, modern economics depends on fossil fuel energy, and reducing this dependence and greenhouse gas (GHG) emissions is likely to increase on the severity of the problem, often disagree on the best approach to solve the problem. Climate change is a global problem and can be solved through international cooperation. The CIA and the Department of Defence both say that climate change is one of the biggest security issues facing this planet. Instability can cause international conflict. The United States must lead the world by working with China, Russia, India and the rest of the International Community to transform our energy system to move away from fossil fuel towards energy efficiency and sustainable energy. We need a global commitment to reduce carbon pollution emission. The US needs to lead the international community in the light against climate change to maintain American economic strength and global security. The United States has a unique opportunity to lead the international community in innovating strategies to cut climate pollution and avoid the most devastation of global climate change and adapt to the impacts that we cannot avoid. Our progress in reducing pollution levels over the last ten years has given us the credibility to demand that major developing nations also take bold action. In the past years, we have seen historic new commitments from countries like China and India who are for the first time putting forth their own aggressive climate agendas. The US and congressional interests in India cover a wide spectrum of issues, ranging from the militarized dispute with Pakistan and weapons proliferation to concerns about regional security, terrorism, human rights, and health, energy, trade and investment opportunities. In the 1990s, India-US relations were particularly affected by the demise of the Soviet Union-India's main trading partner and most reliable source of economic and military assistance for most of the Cold War and New Delhi's resulting need to diversify its international relationships. Also significant were India's adoption of significant economic policy reforms beginning in 1991, a deepening bitterness between India and Pakistan over Kashmir, and signs of a growing Indian preoccupation with China as a potential long-term strategic rival. With the fading of Cold War constraints, the United States and India began exploring the possibilities for a more normalized relationship between the world's two largest democracies. Throughout the 1990s, however, regional rivalries, separatist tendencies, and sectarian tensions continued to divert India's attention and resources from economic and social development. Fallout from these unresolved problems particularly nuclear proliferation and human rights issues presented major irritants in bilateral relations. India's May 1998 nuclear tests were an unwelcome surprise and seen to be a policy failure in Washington, and they spurred then-Deputy Secretary of State Strobe Talbott to launch a series of meetings with Indian External Affairs Minister Jaswant Singh in an effort to bring New Delhi more in line with US arms control and non-proliferation goals.

While this proximate purpose went unfulfilled, the two officials soon engaged a broader agenda on the entire scope of US-India relations, eventually meeting fourteen times in seven different countries over a two-year period. The Talbott-Singh talks were considered the most extensive US-India engagement up to that time and likely enabled circumstances in which the United States could play a key role in defusing the 1999 Kargil crisis, as well as laying the groundwork for a landmark US presidential visit in 2000. President Bill Clinton's March 2000 visit to South Asia seemed a major US initiative to improve relations with India. One outcome was a Joint Statement in which the two countries pledged to "deepen the India-American partnership intangible ways". "A US-India Joint Working Group on Counterterrorism" was established that year and continues to meet regularly. During his subsequent visit to the United States later in 2000, Prime Minister Atal Bihari Vajpayee addressed a joint session of Congress and issued a second Joint Statement with President Clinton agreeing to cooperate on arms control, terrorism, and HIV/AIDS. In the wake of the September, 2001 terrorist attacks on the United States, India took the immediate and unprecedented step of offering to the United States full cooperation and the use of India's bases for counterterrorism operations. Engagement was accelerated after a November, 2001 meeting between President Bush and Prime Minister Vajpayee, when the two leaders agreed to greatly expand US-India cooperation on a wide range of issues, including regional security, space and scientific collaboration, civilian nuclear safety, and broadened economic ties. Notable progress has come in the area of security cooperation, with an increasing focus on counterterrorism, joint military exercises, and arms sales.

In late 2001, the US-India Defence Policy Group met in New Delhi for the first time since India's 1998 nuclear tests and outlined a defence partnership based on regular and high-level policy dialogue. Prime Minister Manmohan Singh paid a landmark July 2005 visit to Washington, where what may be the most significant joint US-India statement to date was issued. In March 2006, President Bush spent three days in India, discussed further strengthening a bilateral "global partnership", and issued another Joint Statement. The Bush Administration has vowed to "help India become a major world power in the 21st century," and US-India relations are today conducted under the rubric of three major "dialogue" areas: strategic (including global issues and defence), economic (including trade, finance, commerce, and environment), and energy. President Bush's 2002 National Security Strategy of the United States stated that "US interests require a strong relationship with India." The 2006 version claims that "India now is poised to shoulder global obligations in cooperation with the United States in a way befitting a major power." In the course of an annual assessment of global threats, Director of National Intelligence Mike McConnell said, we expect India's growing confidence on the world stage as a result of its sustained high rates of economic growth will make New Delhi a more effective partner of the United States but also a more formidable interlock tor in areas of disagreement, particularly in the WTO.

In late 2007, Under Secretary of State Burns, who travelled to New Delhi at least eight times over a two-year period, penned an article lauding stronger US-India relations while also identifying "considerable hurdles" to achieving an effective "global partnership." Foremost among these were terrorism, narcotics trafficking, and nuclear proliferation, whichever must be dealt with through stronger bilateral security ties that will include defence sales. At a 2007 US-India business conference in Washington, Secretary of State Rice laid out the perspective that, We in America look to the rise of India as an opportunity, a chance to work with a great fellow democracy to share not only the benefits of the international system, but indeed, the burdens and the responsibilities of maintaining it, of strengthening it, and defending it. We are eager to continue charting a global partnership with India, one that addresses the global challenges upon which the safety and success of every nation now depends: stemming nuclear proliferation, fighting terrorism, combating disease, protecting the environment, supporting education and upward mobility, expanding economic development, and promoting freedom under the rule of law.

Some analysts, however, see great potential but little likelihood of India becoming a major global power in the foreseeable future. Despite possession of a large, youthful, entrepreneurial population, a booming national economy, and growing power projection capabilities in the military realm, there remains much doubt about the capacity of India's leaders to engage in effective long-term strategic thinking and policy making. One senior Washington-based India-watcher has opined that, "The Indian strategic community is hopelessly strategic," and that its political community is "too domestically focused," thus precluding India's emergence as a substantive major power. Some Indian analysts agree that their national leaders lack a "coherent national grand strategy." Recognition of India's increasing stature and importance and of the growing political influence some 2.3 million Indian-Americans is found in the US Congress, where the India and Indian-American Caucus is now the largest of all country-specific caucuses. Over the past seven years, legal Indian immigrants have come to the United States at a more rapid rate than any other group. In 2005 and 2006, the Indian-American community, relatively wealthy, geographically dispersed, and well-entrenched in several US business sectors, conducted a major (and apparently successful) lobbying effort to encourage congressional passage of legislation to enable US-India civil nuclear cooperation.

6.2 Historical Perspective of India-US Relations

Since India's independence in 1947, ties with the United States have weathered Cold War-era distrust and estrangement over India's nuclear program. Relations have warmed in recent years and cooperation has strengthened across a range of economic and political areas. On August 15, 1947, Britain declared the end of its colonial rule of the subcontinent and passes the Indian Independence Act, which divides the territory into Muslim-majority Pakistan and secular India, whose population is majority Hindu. Violent clashes between Hindus, Sikhs, and Muslims follow, and as many as a million die in bloodshed amidst the forced migration of up to twenty million people.

October 13, 1949: Prime Minister Nehru Visits US

Prime Minister Pandit Jawaharlal Nehru meets US president Harry S. Truman on a multi-week tour of the United States. The trip precedes India's formal proclamation of neutrality in the developing Cold War, in which it would take a leadership role within the Non-Alignment movement. This sets the tone for US-India relations throughout the Cold War, creating constraints within the relationship, as well as opportunity for amity between Delhi and Moscow.

December 9, 1959: President Eisenhower Visits India

President Dwight Eisenhower is the first serving US president to visit the country. Eisenhower meets with President Rajendra Prasad and Prime Minister Jawaharlal Nehru, and addresses Parliament

October 20, 1962: India, China Fight Border War

War breaks out between India and China over a disputed frontier. Prime Minister Jawaharlal Nehru writes to President John F. Kennedy to request support from the United States. Washington supports India in the conflict, recognizing the McMahon line as the border, and provides air assistance and arms. Until the 1965 Indo-Pakistan War, strategic and military ties between Washington and Delhi remain close.

1963: USA Agronomist Spurs Food Revolution

Norman Borlaug travels to India to begin testing high-yield wheat varieties. His collaboration with Indian scientist Dr. M.S. Swaminathan results in the "Green Revolution," and India goes from food scarcity to self-sufficiency within a decade.

1971: India, Pakistan Go to War

India and Pakistan become embroiled in their third conflict as Pakistan descends into a civil war that ends with the creation of Bangladesh, formerly East Pakistan, on December 6. Despite evidence of the Pakistan Army's violence against its own citizens in East Pakistan, the United States sides with Islamabad, given its mediating role in Nixon's rapprochement with China. India also signs a twenty-year Treaty of Friendship and Cooperation with the Soviet Union in August, sharply deviating from its previous position of non-alignment in the Cold War.

May 18, 1974: India Completes First Nuclear Test

India detonates its first nuclear device, becoming the first nation outside the five permanent members of the United Nations Security Council to have declared nuclear capabilities. The move contributes to a period of estrangement between the United States and India that lasts over two decades.

January 1, 1978: President Carter Visits India

The US President Jimmy Carter visits India on a three-day official trip to meet with Indian president Neelam Sanjiva Reddy and Prime Minister Morarji Desai, and address Parliament. Desai reciprocates with an official six-day visit to Washington in June.

March 10, 1978: U.S. Enacts Non-proliferation Act

The Carter administration enacts the Nuclear Non-proliferation Act, which requires countries not included in the Non-proliferation Treaty which includes India—to allow inspections of all nuclear facilities by the International Atomic Energy Agency. India refuses, and Washington ends all nuclear assistance to Delhi.

October 31, 1984: Indira Gandhi Assassinated

Prime Minister Indira Gandhi is assassinated at her New Delhi residence by Sikh security guards seeking vengeance after her authorization of a military attack on a revered Sikh temple in Amritsar five months prior. Gandhi, the daughter of India's first Prime Minister, Jawaharlal Nehru, first took office in 1966. She made a state visit to Washington in 1982, which was followed by a high-level US visit led by Vice President George H.W. Bush. She is succeeded by her son, Rajiv.

December 3, 1984: Bhopal Leak

A toxic gas and chemical leak at American-owned Union Carbide Pesticide Plant in Bhopal, India kills thousands. India unsuccessfully seeks extradition of the company's chief executive from the United States for criminal prosecution as the death and disability toll climbs to the tens of thousands in the ensuing years. The incident harms US-India relations, and continues to complicate the bilateral relationship years after.

May 20, 1990: US Crisis Mission to Region

Deputy National Security Adviser Robert Gates travels to India and Pakistan to defuse tensions over the rapidly escalating insurgency in Kashmir. The trip comes amid fears of potential nuclear warfare between Pakistan and India.

July 24, 1991: Economic Reforms

The government of Prime Minister P.V. Narasimha Rao, launches sweeping economic reforms that help expand economic ties with the United States. Finance minister Manmohan Singh oversees the opening of India's economy to international trade and investment, deregulation, initiation of privatisation, tax reforms, and inflation-controlling measures that catalyse decades of fast growth.

May 11, 1998: India Tests Nuclear Devices

The Indian government announces the completion of a series of underground nuclear tests close to the border with Pakistan, surprising US intelligence organizations and raising fears the move could spark a regional nuclear arms race. The tests draw international condemnation and badly damage India's relationship with the United States. After recalling the US ambassador to India, President Bill Clinton imposes economic sanctions, required under US law.

May 3, 1999: Pakistan, India Clash in Kashmir

Pakistani forces infiltrate Indian-administered Kashmir. India launches air strikes in return, and armed conflict continues through early July. After President Clinton summons Pakistani Prime Minister Nawaz Sharif to Washington for a Fourth of July emergency meeting, Sharif withdraws Pakistani forces from their positions beyond the Line of Control.

March 20, 2000: Clinton Trip Signals Warming Ties

President Bill Clinton makes the first US presidential trip to India since 1978. The visit ends the estrangement of the post-1998 Indian nuclear weapons tests, although the Clinton administration presses India's government to sign the Comprehensive Test Ban Treaty. The Indo-US Science and Technology Forum is also established during the visit. As India's economy begins to take off, the trip indicates a further shift in Washington's regional orientation away from its Cold War alliance with Pakistan.

September 22, 2001: US Lifts India Sanctions

The George W. Bush administration lifts all remaining US sanctions that were imposed on India after its 1998 nuclear test. Most economic sanctions had been eased within a few months of their imposition, and Congress authorized the president to remove all remaining restrictions in 1999.

March 15, 2005: Energy Security Dialogue

US secretary of state Condoleezza Rice visits New Delhi, where she and Indian officials agree to start a dialogue on energy security. The visit underscores an upswing in relations despite tensions over India's possible energy cooperation with Iran and the US sale of fighter jets to Pakistan.

June 28, 2005: US-India Sign New Defence Framework

The United States and India sign the New Framework for the US-India Defence Relationship, which sets priorities for defence cooperation in maritime security, humanitarian assistance/disaster relief, and counterterrorism. In October, the two countries conducted the largest naval exercise to date, followed by major air and land exercises.

July 18, 2005: Landmark Civil Nuclear Deal

India and the United States ink the Civil Nuclear Cooperation Initiative, a ten-year defence framework that lifts a three-decade US moratorium on nuclear trade with India. Under the agreement, India agrees to separate its civil and military nuclear facilities and place all its civil resources under IAEA safeguards. In exchange, the United States agrees to work toward full civil nuclear cooperation with India. Congress gave final approval in October 2008.

March 1, 2006: President George W. Bush Visits India

US president George W. Bush makes a visit to India, where he and Indian Prime Minister Manmohan Singh finalize the framework of the civil nuclear deal and boost security and economic ties. The nuclear deal, completed in July 2007, makes India the only country outside of the Non-proliferation Treaty that has nuclear capabilities and is allowed to participate in nuclear commerce.

November 26, 2008: Terrorists Attack Taj Mahal Palace Hotel

Lashkar-e-Taiba terrorists from Pakistan attack the Taj Mahal Palace Hotel in Mumbai. More than 300 citizens die in the three-day conflagration, including six Americans. The United States cooperates closely with Indian authorities, sending FBI investigators and forensics experts.

November 2009: Prime Minister Manmohan Singh Begins US State Visit

US president Barack Obama hosts the inaugural state visit of Indian Prime Minister Manmohan Singh. Despite its symbolic importance, the trip fails to yield any significant breakthroughs in the bilateral relationship.

April 5, 2010: Economic and Financial Partnership
US Treasury Secretary Timothy Geithner makes his first official trip to India to launch the new US-India Economic and Financial Partnership with Indian finance minister Pranab Mukherjee. The ministerial-level meetings kick off an effort to institutionalize deeper bilateral relations on economic and financial sector issues.

June 1, 2010: US, India Hold First Strategic Dialogue
The United States and India formally convene the first US-India Strategic Dialogue. A large, high-ranking delegation of Indian officials visits Washington, DC, and Secretary Clinton lauds India as "an indispensable partner." President Obama says the relationship "will be a defining partnership in the twenty-first century." Subsequent dialogues follow annually.

November 5, 2010: Obama Backs India Bid for UN Security Council
President Obama visits India, where he addresses Parliament and backs the country's long-held bid for a permanent seat on the United Nations Security Council. The trip also highlights the countries' economic ties, with Obama announcing $14.9 billion in trade deals. However, trade concerns around access to Indian markets and issues surrounding civil nuclear cooperation cloud the talks.

July 19, 2011: US, India Ink Cyber Security Memorandum
The United States and India sign a Memorandum of Understanding in New Delhi to promote closer cyber security cooperation. The agreement is designed to fulfil one of the pillars of the US-India Strategic Dialogue.

May 30, 2012: Panetta Boosts Military Ties
Secretary of Defence Leon Panetta visits India to bolster military ties in the wake of the Obama administration's announced "pivot" to Asia. The trip marks the first such visit since former secretary of defence Robert Gates met with Indian counterparts in January 2010.

September 27, 2013: Singh Makes Last Visit to Washington
Manmohan Singh visits Washington in his last visit to the United States as India's prime minister. The trip, which focuses on security, trade,

immigration reform, and the civilian nuclear deal, marks the third meeting between Singh and Obama in four years. It comes amid a backdrop of domestic political issues in Delhi, a troubled Indian economy, and a government shutdown crippling Washington.

March 31, 2014: Diplomatic Row Sours Ties

The US embassy in India announces Ambassador Nancy Powell's resignation in the wake of a dispute over the arrest of an Indian diplomat in New York. The announcement comes amid the run-up to high-profile national elections.

May 20, 2014: Obama Invites Indian Prime Minister to US

The Hindu nationalist BJP party wins national elections in a landslide, elevating current Indian Prime Minister. President Barack Obama congratulates Indian Prime Minister and invites him to the White House, reversing an earlier visa ban. The current Indian Prime Minister had been barred from entering the country over the concerns of the US about the 2002 massacre of Muslims in the state of Gujarat, which occurred when he was the state's chief minister.

6.2.1 Major Exchanges in 2015

There were a number of high-level delegations in both directions in 2015. In January, Secretary of State John Kerry led the US delegation to the Vibrant Gujarat Summit in Ahmedabad. In February, US Secretary of Treasury Jacob Lew visited India for the fifth meeting of the Economic and Financial Partnership Initiative with our Finance Minister. US Secretary of Transportation Anthony Foxx visited India in April for meeting with his counterpart Ministers in India. Secretary of Defence Ashton Carter paid a bilateral visit to India in June. From the Indian side, Finance Minister and MOS (IC) for Environment,

Forests & Climate Change visited Washington DC in April, the former for the Spring meeting of the IMF/World Bank and the latter for the Major Economies' Forum meeting. Finance Minister again visited the US in June to promote Investment into India. External Affairs Minister and MOS (IC) for Commerce & Industry co-chaired the first meeting of the Strategic and Commercial Dialogue with their US counterparts in

Washington DC in September. MOS (IC) for Power held the Energy Dialogue with his US counterpart in September in Washington DC. MOS (IC) for Power, Coal and New & Renewable Energy also attended the Climate and Clean Energy Investment Forum 2015 hosted by Secretary of State Kerry in Washington DC in October. MOS (IC) for Commerce & Industry held the 9th meeting of the Trade Policy Forum with the US Trade Representative in Washington DC in October. Raksha Mantri visited the US at the invitation of his US counterpart in December 2015. Chief Minister of Haryana led a delegation of the State government in August. There have been numerous visits of Parliamentarians and senior officials in both directions.[1]

- United States Secretary of State John Kerry and Secretary of Commerce Penny Pritzker welcomed India's External Affairs Minister Sushma Swaraj and Minister of State for Commerce and Industry Nirmala Sitharaman for the first US-India Strategic and Commercial Dialogue held in Washington DC on 22 September 2015.
- The sides appreciated the intense engagement between India and the United States under various institutional bilateral dialogue mechanisms and people-to-people contacts. The Sides welcomed India's announcement to hold a regional Pravasi Bhartiya Divas (PBD) in Los Angeles in November 2015, to further increase people-to-people contacts between the two countries.
- Both sides assessed that ties between the United States and India have never been stronger as reflected by unprecedented strategic cooperation, record levels of bilateral trade and investment, and more than 2 million annual visits between their citizens, students, and entrepreneurs.
- The sides pledged to build on this momentum by pursuing new areas of collaboration, leveraging the talents of government and the private sector to make their nations more secure and prosperous.

6.3 India-US Strategic Partnership to Deal with Climate Change

India-US bilateral relations have developed into a "global strategic partnership", based on shared democratic values and increasing convergence of interests on bilateral, regional and global issues. The emphasis placed by the Government in India on development and good governance has created opportunity to reinvigorate bilateral ties and enhance cooperation under the motto *"Chalein Saath Saath"* (Forward Together We Go), and *"Sanjha Prayas, Sabka Vikas"* (Shared Effort, Progress for All) adopted during the summits of our leaders in September 2014 and January 2015 respectively.[2] Regular exchange of high level political visits has provided sustained momentum to bilateral cooperation, while the wide-ranging and ever-expanding dialogue architecture has established a long-term framework for India-US engagement. Today, the India-US bilateral cooperation is broad-based and multi-sectoral, covering trade and investment, defence and security, education, science and technology, cyber security, high-technology, civil nuclear energy, space technology and applications, clean energy, environment, agriculture and health. Vibrant people-to-people interaction and support across the political spectrum in both countries nurture our bilateral relationship.[3] From controversies around legislations back home to comments about rural women, much had been said and written about the US-visit of the then Prime Minister, Dr. Manmohan Singh. Lost in the noise is the substance of the India-US strategic partnership which got a substantial impetus on specific dimensions. Apart from the usual rhetorical flourishes, the US-India joint statement noted considerable progress on the issues of nuclear energy and defence cooperation. But one particular aspect of the partnership that got a major boost and went nearly unnoticed is cooperation on combating climate change.[4]

In 2016, the United States reaffirms its commitment to join the Agreement as soon as possible this year. India similarly has begun its processes to work toward this shared objective. The leaders reiterated their commitment to pursue low greenhouse gas emission development strategies in the pre-2020 period and to develop long-term low greenhouse gas emission development strategies. The United States and

India share common climate and clean energy interests and are close partners in the fight against climate change. Leadership from both countries helped galvanize global action to combat climate change and culminated in the historic Paris Agreement reached last December. Both countries are committed to working together and with others to promote full implementation of the Paris Agreement to address the urgent threats posed by climate change. India and the United States recognize the urgency of climate change and share the goal of enabling entry into force of the Paris Agreement as early as possible.[5]

Climate change has been an interesting topic for India-US relations. While both countries officially recognise the need for urgency in combating climate change, they have been at different poles in international negotiations. The US emphasises that all countries (including developing countries like India) should take on commitments, with transparency and review systems to verify compliance. India on the other hand, along with the so-called 'G-77+China' bloc of developing countries, emphasises that the primary responsibility of taking action rests with developed countries. They argue that the developed countries have created the climate change problem in the first place, that they have a 'historical responsibility' to address it, and that the principle of 'common but differentiated responsibilities' (CBDR) must govern all actions. This effectively means that the primary effort to reduce greenhouse gas (GHG) emissions should be made by developed countries, and to the extent that developing countries are expected to act, they must be given access to appropriate technologies and adequate finance to enable action.[6] These countries are making progress on a bilateral basis.

The US-India joint statement is a significant step forward in this direction. Both the countries have agreed to set up an India-US climate change working group to develop and advance action-oriented cooperation and convene the India-US task force on hydro fluorocarbons (HFCs). India and the US signed the Memorandum of Understanding on Energy Security, Clean Energy and Climate Change in November 2009. According to this MoU, both the countries agreed to cooperate on several issues like strengthening of R&D efforts for use of unconventional sources of cleaner fuels like coal bed methane and shale gas; better management of forest resources, reforestation and reducing carbon emissions resulting from deforestation and change in land use, and cooperation on adaptation measures by studying changing weather patterns.

The US-India strategic partnership has become especially significant in the context of the United States rebalancing policy in the Asia-Pacific region. Recognizing the profound threat of global climate change, the sides stressed the importance of working together and with other countries to conclude an ambitious climate agreement in Paris in December 2015, understanding that meeting this goal will require concerted action by all countries and the international community.[7] The sides looked forward to the early signing of a new five year Memorandum of Understanding on Energy Security, Clean Energy and Climate Change. The sides welcome the productive meeting of the US-India Energy Dialogue in 2015 and looked forward to expanding bilateral engagement in this field and energy efficiency, as well as to develop and exchange information on cleaner fossil energy resources such as unconventional oil and gas and carbon capture sequestration. In particular the two sides welcomed significant progress in implementing President Obama and Indian Prime Minister's commitment to strengthen and expand the highly successful US-India partnership to Advance Energy (PACE), including:

Greening the Grid, a U.S. 30 million, 5-year initiative to scale up renewable energy integration into India's power grid. Two activities to promote off-grid clean energy access include the PACE setter Fund, a joint US$ 41 million in finance for clean energy entrepreneurs and research on smart-grids and energy storage for grid application as the fourth stream under PACE-R. The sides noted conclusion of the first phase of the US-India clean Energy Finance Task Force and in the next phase will deepen engagement through design and implementation in a time bound manner of select pilot projects in catalysing accelerated flows of untapped sources of capital to support India's ambitious clean energy goals. The sides convened the second round of bilateral consultations under the India-US Joint Working Group on Combating Climate Change, continuing their enhanced dialogue on multilateral negotiations and strengthening bilateral efforts in the areas of clean energy, adaptation, forest sector programs, and air quality. Drawing on the academic and scientific cooperation that underpins the US-India partnership, the sides looked forward to launch a new Fulbright-India Climate Fellowship for capacity building in climate research between India and the US Department of State.[8] India, US plan initiative includes cooperation on reducing air pollution in Indian cities under a five-year agreement.

The US-India joint announcement on Climate and clean Energy includes announcements on issues including the Paris Agreement, Montreal Protocol and clean energy. India and the US already have several programmes of cooperation, including the US-India Partnership to Advance clean Energy, and both sides have agreed to expand policy dialogues and technical work on low greenhouse gas emissions technologies. They also decided to discuss ways to reduce the environmental and emissions impact of heavy vehicles and auto fuels by working to adopt cleaner fuels and emission standards in India. Demonstrating support for the ISA, the United States and India plan to jointly launch an effort to promote access to off-grid solar power. The Government of India has been a strong partner in the Clean Energy Ministerial (CEM), a forum of the world's major economics, working together to drive the transition to clean energy. India has demonstrated how international collaboration such as efforts to establish lighting test standards through the CEM's appliance efficiency initiative, can lead to huge benefits. India's domestic efficient lighting program, which has deployed over 100 million efficient LED, inspired the CEM's Global Lighting Challenge. The other major area of close cooperation would be hydrofluorocarbons (HFCs), as the two countries agreed to build on "prior understanding from September 2014" concerning the phase down of HFCs. Both Obama and Prime Minister Indian Prime Minister decided to cooperate on making concrete progress in the Montreal Protocol later this year.

These are the few areas where both the governments have been working-

- A Strengthened and Expanded US-India Partnership to Advance Clean Energy.

- Accelerating Renewable Energy Deployment
- Grid integration
- Solar Resource Data
- US India clean Energy Finance Task Force
- New Clean Energy Finance initiative
- US India Catalytic Solar Finance Program
- Energy Efficiency: "Expanding on historical collaboration, the United States and India are updating the Energy Conservation Building code and supporting comprehensive state-level policies for the current framework, while making significant progress on space cooling collaboration."[9]
- Innovation : President Obama and Prime Minister Indian Prime Minister launched Mission Innovation at the 21st Conference of the parties (COP-21) to the United Nations Framework Convention on Climate change in Paris, pledging along with 18 other countries, to seek to double clean energy research and development investment over five years. The two countries announced a $30 million public-private research effort in small grid and grid storage.

On 31 August 2016, India and the United states held the second India-US Strategic and Commercial Dialogue in New Delhi. The then External Affairs Minister of India, Sushma Swaraj, and Minister of State for Commerce and Industry, Nirmala Sitharaman, co-chaired, the dialogue with the US Secretary of State John F. Kerry and US Secretary of Commerce Penny Pritzker. The sides expressed deep satisfaction at the expanded and strengthened course of bilateral engagement in recent times and reiterated their shared commitment to advance mutual prosperity, global peace, and stability. Climate Change has been at the core of the European Union's (EU) environmental policy since 1988 and has been integral to its pursuit of international leadership to foster multilateral Climate protection. To understand whether and how climate change links with the peace and security agenda, it is necessary first to examine how relevant actors are framing the issue. These emerging challenges are the long-term security implications of climate change, particularly in the vast and vulnerable Asia-Pacific region, where the nexus of humanity and the effects of climate change are expected to be most profound. These changes are altering the Asia-Pacific security environment and to revaluate and adjust our long-term "whole of government" strategic priorities and approaches in the region. For more than seven decades the United States, in coordination with our allies and partners, effectively shaped and oversaw the post-World War II Asia-Pacific security environment.

Through an intricate network of alliances and strategic partnerships, backed by strong diplomacy, productive economic policy and a robust and largely uncontested forward military posture, the United States created and maintained a security environment that history will agree dealt effectively with the traditional Security challenges and underpinned the significant social, political and economic success that can be observed today in many parts of the Asia-Pacific region. As we look forward, it is without question in our best interest to see this success continue. However, success will require planning and preparation for the full range of the Asia-Pacific's traditional and non-traditional security challenges, which must include what may prove to be the Asia-Pacific's greatest long-term challenges, i.e. climate change. 'Promoting Energy Access through Clean Energy' (PEACE) to "rapidly deploy innovative renewable energy solutions to those who currently lack access to energy", this is a significant step forward by both counties.

Nevertheless, these three concrete elements demonstrate a new openness on the part of the two countries to engage on what the leaders of the two countries have themselves called "a defining global challenge of our times". The task ahead will be to maintain the momentum and ensure that we 'walk the talk,' showing results in a clear time-bound manner, now that a window has been opened.[10] Over the past 15 years, three significant developments have helped bring about the recent dramatic strengthening of US-India ties. First, the end of the Cold War removed the US-Soviet rivalry as the principal focus of US foreign relations and the rationale for India's nonalignment policy. Second, India's historic economic reforms of the early 1990's, led by Manmohan Singh, then finance minister and later prime minister, opened India to the global economy for the first time and catalysed the extraordinary boom in private-sector trade and investment between the United States and India that continues today. Under Trump's presidentship, current Indian Prime Minister, was invited to the white house for close interaction despite having sharp disagreements with India over issues like trade, climate change, and H-1B visa. Such issues got side-lined during the summit of two leaders and released a joint statement emphasising on strengthening defence partnership, cooperating on counterterrorism efforts and boosting economic ties.[11] Finally, as the twenty-first century began, the global order started to undergo a tectonic shift, and India's emergence as a global force was obvious for all to see.[12] That this new US-India partnership is supported by a bipartisan consensus in both countries considerably strengthens the prospects for its success. In India, both the ruling Indian National Congress and the opposition Bharatiya Janata Party have worked for over a decade to elevate India's ties with the United States. In the United States, shortly after the beginning of India's economic liberalization, President Clinton signalled Washington's desire to forge a new era of commerce and investment between the two countries. And after India's May 1998 nuclear tests, then Deputy Secretary of State Strobe Talbott engaged India's then foreign minister, Jaswant Singh, in 14 rounds of talks over two and a half years. Talbott's negotiations with Singh were Washington's first truly sustained strategic engagement with the Indian leadership.[13]

United States can be called as hypocritical for seeking a strategic partnership with India while simultaneously targeting it with punitive sanctions. The United States abandoned this approach during George W. Bush's first term and had completed a total reversal by the end of his second, when the civil nuclear agreement lifted restrictions on nuclear commerce. Although negotiated by a Republican administration, the deal passed a Democratic-controlled Congress with overwhelming bipartisan backing (including the votes of the Senators Joe Biden, Hillary Clinton, and Obama), demonstrating that support for strengthened US-Indian relations extends across party lines in Washington, much as it does in India. Also bolstering the relationship are both American and Indian business people, the Council on Foreign Relations 2010 shows that nearly 100,000 Indian students currently studying in US schools, some three million Indian Americans, and the tens of millions of Indians with relatives in the United States.[14] There is a broad perception in India's strategic community that despite the many new elements of US-Indian cooperation, the United States has recently been tilting toward Pakistan by ramping up its aid to the country and its military-to-military cooperation with Pakistan. Particularly concerning to many Indians is the suggestion made by some influential US commentators that Washington should push New Delhi to alter its military posture toward Pakistan? Or even make lasting concessions on Kashmir? US is in the hope that Pakistan would then remove resources from its eastern border and focus them instead on fighting al-Qaeda and the Taliban.[15] The United States and India share important interests: both seek to restore global growth, protect the global commons, enhance global energy security, and ensure a balance of power in Asia. They must therefore increase the scope, quality, and intensity of their cooperation at every level. But the ultimate test of their relationship will be whether Washington and New Delhi can turn their common interests into complementary policies around the world.[16] President Obama and Indian Prime Minister committed in September 2014 to a new and enhanced strategic partnership on energy security, clean energy, and climate change.

Since that time, the United States and India have made important strides together to advance this strategic partnership, and to our related cooperation on environmental stewardship. There is a tremendous strategic upside to our growing engagement with India. That is why building a close US-India partnership should be one of the United States' highest priorities for the future. It is a unique opportunity with real promise for the global balance of power. Today we share an abundance of political, economic, and military interests with India. We share an attachment to democracy and individual rights founded on an instinctive mistrust of authoritarianism. And in an age of anti-Americanism, according to the most recent Pew Global Attitudes survey, nearly six in ten Indians view the United States favourably. President Obama visited India on 25-27 January 2015, and during the visit the two sides issued a Delhi Declaration of Friendship and adopted a Joint Strategic Vision for Asia-Pacific and the Indian Ocean Region. Both sides elevated the Strategic Dialogue between their Foreign Ministers to Strategic and Commercial Dialogue of Foreign and Commerce Ministers. Indian Prime

Minister again visited the US on 23-28 September 2015, during which he held a bilateral meeting with President Obama, interacted with leaders of business, media, academia, the provincial leaders and the Indian community, including during his travel to the Silicon Valley.

- **US Strongly Supports India's Nuke Club NSG Bid:** The US "strongly supports" India's NSG membership bid, President Barack Obama told Indian Prime Minister as they discussed the immediate priorities in the strategic partnership, including deepening the civil nuclear cooperation and combating climate change.
- **US Senate Move To Give India Global Strategic Partner Status**: Asserting that the US and India face mutual security threats and a robust defence partnership is in the interest of both countries, a top Republican Senator had moved a legislation asking the President to recognise India as America's "global strategic and defence partner."
- **Deeper Cooperation with India on Nuclear Issues:** The United States would like to see "deeper bilateral cooperation" with India on nuclear issues, the White House had said, ahead of the Nuclear Security Summit in Washington. The first Nuclear Security Summit was held in Washington, DC in 2010 and was followed by additional Summits in Seoul in 2012 and Hague in 2014 and Washington D.C. in 2016 which was attended by Indian Prime Minister.
- These Summits have achieved tangible improvements in the security of nuclear materials and stronger international institutions that support nuclear security.[17]
- **2015, A Transformative Year For India-US Relationship**: In the annals of Indo-US ties, 2015 would be remembered as a transformative year which set the tone for a new dimension of active engagements at the top leadership level and a growing defence partnership that is set to become an anchor of global security.
- **US Supports India's Membership to Missile Technology Control Regime**: The US has supported India's entry into the

Missile Technology Control Regime (MTCR) that would enable it to share sensitive missile technologies like armed drone with New Delhi.

- **'Our Goal is to Be India's Best Friend', Says US Vice President Joe Biden**: India and the United States kicked off a crucial dialogue in Washington with the US leaders saying America wants to be "India's best friend" and India asserting that their relationship is defined by natural synergy of democracies.

6.3.1 Post-2015, India-US Strategic tie up on Climate Change

India has been sourcing liquified natural gas (LNG) and crude oil from the US, with Indian companies investing $4 billion in US shale gas assets. India is now the fourth-largest international market for the US crude oil and the fifth largest for the US liquefied natural gas. Indian firms have also concluded several contracts for sourcing crude from the United States and are expanding their investments in the US energy sector, creating jobs and economic opportunities.[18]

- With India's energy demand expected to grow at 4.2 per cent per year over the next 25 years, it has contracted 9 million metric tonnes per annum (mmtpa) of LNG from the US, making it the sixth-largest buyer of US LNG.
- In a first, Indian Oil Corp., the country's largest refiner, also inked two-term contracts totalling 4.6 million tonnes (mt) of US crude oil for 2019-20 from Norway's Equinor ASA and Algeria's state energy company Sonatrach.
- Tellurian Inc. signed a $7.5 billion agreement for India's Petronet LNG Ltd to buy a stake in its proposed LNG terminal near Lake Charles, Louisiana, in what could potentially be one of the largest foreign investments in the US for shipping shale gas abroad. India is the world's fourth-largest LNG importer.
- US Department of Energy (DOE) and the Government of India established the Joint Clean Energy Research and Development Center (JCERDC) to promote clean energy innovations, with a total joint committed funding from both Governments of US$ 50 million.

Four working groups were created under the Strategic Energy Partnership oil and gas, power and energy efficiency, renewable energy and sustainable development. India and the US are advancing cooperation and dialogue on climate change through a high-level Climate Change Working Group and a Joint Working Group on Hydrofluorocarbon. A new US-India Partnership for Climate Resilience has been agreed to, in order to advance capacity for climate adaptation planning, as also a new US-India Climate Fellowship Program to build long-term capacity to address climate change-related issues. In 2016, the two sides finalized a package to provide concessional finance to support clean energy projects, coordinate efforts on clean energy investment and provide liquidity to small scale renewable energy investors. Civil nuclear energy partnership and collaboration between Nuclear Power Corporation of India Limited (NPCIL) and Westinghouse Electric Company for the establishment of six nuclear power plants in India. US oil exports to India have been surging in recent years from zero in 2016 to nearly ten million barrels in 2017, to 48 million barrels in 2018, to potentially 100 million barrels in 2019.

6.4 Strategic Cooperation in Global Issues

There have been regular contacts at political and official levels on bilateral, regional and global issues. India and US in the recent years instituted structured dialogues covering East Asia, Central Asia, West Asia, Africa and the Indian Ocean Region. India and the US also have a trilateral with Japan (the first Ministerial-level meeting of the Foreign Ministers took place on 29 September 2015 in New York) and a trilateral with Afghanistan (last meeting held in 2013). Matters relating to international security and disarmament, multilateral export control regimes are reviewed under the Strategic Security Dialogue, while issues relating to high-technology trade are discussed in the India-US High Technology Cooperation Group (HTCG); both these groups, led by Foreign Secretary of India and her counterparts in the US Departments of State and Commerce respectively, last met in November 2014 in New Delhi. The two sides have agreed to work closely for India's phased entry into the global export control regimes to strengthen global non-proliferation, arms control, as well as nuclear security.[19]

- The Sides recognized that the India-US partnership was a significant contributor to the peace, stability and prosperity in the Indian Ocean and Asia-Pacific regions and around the globe.
- Building on successful cooperation in Asia, the Sides welcomed continued cooperation under the Joint Strategic Vision for the Asia-Pacific and Indian Ocean Region agreed by President Obama and Indian Prime Minister.
- The US Side commended India's leadership in evacuating foreign nationals including US citizens, from conflict in Yemen, as well as cooperation between the United States and India on providing earthquake relief in Nepal. The Sides resolved that India and the United States would work as partners in responding to the needs of civilians in global crises.
- Recognizing the centrality of peacekeeping to the UN's efforts for maintenance of international peace and security, the Sides committed to enhance cooperation in peacekeeping capacity building in third countries with a focus on training aspects for UN peacekeepers, especially in identified African countries.
- The Sides announced that the first ministerial meeting of the India-United States-Japan trilateral would convene on the margins of the UN General Assembly in September 2015. They also welcomed Japan's participation in the MALABAR 2015 naval exercise.
- The US Side affirms its support for India's membership in the Missile Technology Control Regime at its upcoming plenary, the Nuclear Suppliers Group, and in the other global non-proliferation export control regimes.
- The Sides pledged to continue high level consultations on Afghanistan, making clear the enduring commitment of India and United States to the Afghan people. The Sides agreed that a sovereign, independent and prosperous Afghanistan is in the interest of peace and security in the region, and will contribute to the global efforts to combat terrorism and extremism.

- The US Side reaffirmed its support for a reformed UN Security Council with India as a permanent member. Both sides committed to ensuring that the Security Council continues to play an effective role in maintaining international peace and security as envisioned in the UN Charter. Both sides are committed to continued engagement on Security Council reform in the UN Intergovernmental Negotiations (IGN) on Security Council Reform.

6.4.1 Defence and Security Cooperation

Defence relationship has emerged as a major pillar of India-US strategic partnership with the signing of 'New Framework for India-US Defence Relations' in 2005 and the resulting intensification in defence trade, joint exercises, personnel exchanges, collaboration and cooperation in maritime security and counter-piracy, and exchanges between each of the three services. The Defence Framework Agreement was updated and renewed for another 10 years in June 2015. The two countries now conduct more bilateral exercises with each other than they do with any other country. An Indian Navy ship took part in Rim of the Pacific (RIMPAC) exercise in 2014 for the first time. Bilateral dialogue mechanisms in the field of defence include Defence Policy Group (DPG), Defence Joint Working Group (DJWG), Defence Procurement and Production Group (DPPG), Senior Technology Security Group (STSG), Joint Technical Group (JTG), Military Cooperation Group (MCG), and Service-to-Service Executive Steering Groups (ESGs).[20]

- Building on the conclusion of their 10-year Defence Framework Agreement, both sides expressed satisfaction with progress on Defence Technology and Trade Initiative pathfinder projects, including the establishment of the Working Groups on Aircraft Carrier Technology and Jet Engine, and the growing cooperation between US and Indian defence industries through the "Make in India" initiative. They looked forward to an early visit of Indian Minister of Defence Parrikar to the United States to deepen discussions on military exercises, defence trade and technology, and areas of cooperation.
- Honouring the memory of victims of terrorism from around the

globe, including citizens of India and United States on 26/11 and on 9/11, the Sides resolved to deepen cooperation to prevent terrorism, endorsing a Joint Declaration on Combating Terrorism to expand our partnership.

6.4.2 Economic Growth

- Building on the success of the 2014 US-India Commercial Dialogue on corrosion control, the Sides agreed to continue this public-private collaboration. They recognized India's work to launch a National Mission on Corrosion Control Technologies and Standards.
- The Sides committed to accelerating progress in infrastructure collaboration. Both sides welcomed the start of Smart City master planning activities in Vizag led by a US private sector consortium. Building on this momentum, the US Side looked forward to working with the Indian Ministry of Urban Development on a Smart Solutions for Smart Cities Reverse Trade Mission visit to the United States coordinated by the US Trade and Development Agency. India welcomed a Smart Cities Infrastructure Business Development Mission in February 2016 led by the Deputy Secretary of Commerce.
- Reflecting that the strength of a nation depends on the health of its people, the Sides reviewed progress from the first meeting of the US-Health Dialogue, welcoming recent cooperative agreements to enhance cooperation in the field of cancer research, prevention, control and management; environmental and occupational health and injuries prevention and control; and LOI on research on antimicrobial resistance. It was agreed that both sides will work together to consider signing a MoU in the field of Mental Health and a MoU between the Ministry of AYUSH and HHS on collaboration in various aspects of Traditional Medicine, including regulatory and capacity building. AYUSH will be organizing a joint workshop with NIH in early 2016 in India to discuss research collaboration on traditional medicine. The first India-US Workshop on Traditional Medicine was held in New Delhi, India, on March 3 and 4, 2016,

as a collaboration between the Ministry of Ayurveda, Yoga and Naturopathy, Unani, Siddha, and Homoeopathy (AYUSH) of the Government of India, the US National Cancer Institute (NCI), National Institutes of Health, and the Office of Global Affairs, US Department of Health and Human Services.

- The Sides highlighted the global effort led by India, the United States, and other partners, to end preventable maternal and child deaths. They hailed the Delhi Declaration signed by 22 nations in August 2015.
- US Side affirmed that it stands ready to assist in India's ambitious goal of providing skills training to 400 million people over the next decade. In anticipation of the next Higher Education Dialogue, the US side intends to create new programs to build capacity for curriculum and teacher development. They also noted the need to continue efforts to implement the decisions taken during previous summits of leaders of the two countries including the knowledge partnership for supporting the IIT Gandhinagar through USAID and in India's Global initiative for Academic Network (GIAN).
- The Sides expressed satisfaction at the progress made on Science, Technology and Innovation collaboration after the meeting of India-US Joint Commission on Science & Technology in November 2014 in New Delhi. A new working group on Agriculture Biotechnology was established to promote Agriculture Science &Technology Research collaboration led by India's Department of Biotechnology and US Department of Agriculture. Under 'Discovery Science' India's Department of Atomic Energy and US Department of Energy are working together to develop a High Intensity Superconducting Proton Accelerator (HISPA). Under the US-India Endowment Fund, prototypes of several innovative technologies have been jointly developed with potential for societal use and commercial application.

6.4.3 Energy and Environment

- Recognizing the profound threat of global climate change, the Sides stressed the importance of working together and with other countries to conclude an ambitious climate agreement in Paris in December 2015, understanding that meeting this goal will require concerted action by all countries and the international community.
- The Sides looked forward to the early signing of a new five year Memorandum of Understanding on Energy Security, Clean Energy and Climate Change. The Sides welcomed the productive meeting of the US-India Energy Dialogue and looked forward to expanding bilateral engagement in this field including in the field of renewable energy and energy efficiency, as well as to develop and exchange information on cleaner fossil energy resources such as unconventional oil and gas and carbon capture sequestration. In particular, the two sides welcomed significant progress in implementing President Obama and Indian Prime Minister's commitment to strengthen and expand the highly successful US-India Partnership to Advance Clean Energy (PACE), which includes the following:
  - Greening the Grid, a US $30 million, 5-year initiative to scale up renewable energy integration into India's power grid.
  - Two activities to promote off-grid clean energy access: the PACE setter Fund, a joint US $7.9 million fund for innovative off-grid clean energy projects and a new public-private partnership that will work to mobilize US $41 million in finance for clean energy entrepreneurs.
  - Research on smart grids and energy storage for grid application as the fourth stream under PACE – R.
- The Sides noted conclusion of the first phase of the US-India Clean Energy Finance Task Force and in the next phase will deepen engagement through design and implementation in a time-bound manner of select pilot projects in catalysing accelerated flows of untapped sources of capital to support India's ambitious clean energy goals.

- The Sides convened the second round of bilateral consultations under the India-US Joint Working Group on Combating Climate Change, continuing their enhanced dialogue on multilateral negotiations and strengthening bilateral efforts in the areas of clean energy, adaptation, forest sector programs, and air quality.
- Drawing on the academic and scientific cooperation that underpins the US-India partnership, the Sides looked forward to launch a new Fulbright-India Climate Fellowship for capacity building in climate research between India's Department of Science and Technology and the US Department of State.
- Noting the importance of preserving the environment and sustaining diverse ecosystems, and acknowledging the rise of international criminal networks engaged in wildlife trafficking, the Sides appreciated finalization of an MOU to enhance their cooperation on conservation and combat wildlife trafficking. The US Side offered support to India's Project Tiger designed to protect the population of Bengal tigers in their natural habitats. The Sides agreed to work together to use the latest technology to combat poaching and protect tigers in India.
- Rooted in shared democratic values and sustained by the ties between Indians and Americans that have strengthened and enriched both nations, both sides affirmed at the close of the dialogue that together the United States and India would continue to expand and deepen their engagement, to build a defining partnership for the 21st century.
- Minister Swaraj, on behalf of the Indian delegation, thanked Secretary Kerry and the US delegation for hosting the first Strategic and Commercial Dialogue in Washington DC. It was agreed to convene the next round in India in 2016.

The "strategic partnership" between the US and India has been actively projected as a multi-faceted, truly all-embracing relationship. There have been many false dawns before in the uncertain evolution of the relationship between India and the United States. If India and America were estranged democracies during the Cold War, they certainly became more engaged republic since the 1990s. Yet, repeated attempts at elevating it into a genuine strategic partnership seemed to end up nowhere. But India's Prime Minister and US President Barack Obama ended their second summit meeting in less than four months by proclaiming that a new chapter has begun in bilateral relations. This time around there is good reason to be a lot less skeptical about the ambitions of two leaders to change the nature and direction of India-US relations. The United States appears careful not to be confrontational with India ahead of the Paris climate conference, after New Delhi reacted sharply to Secretary of State John Kerry's statement earlier this month that India would be a "challenge" in negotiations. On energy and climate change, Indian Prime Minister and President Obama agreed to a new strategic partnership, which expands the 'US-India Partnership to Advance Clean Energy (PACE)' through priority initiatives, including: the 'Energy Smart Cities Partnership' promoting efficient urban energy infrastructure; a program to scale up grid integration of renewable energy in India; and an expansion of the 'Promoting Energy Access through Clean Energy (PEACE)' programme aimed at unlocking private sector investment and accelerating the adoption of cost-effective, super-efficient appliances. In relation to climate change, the two leaders pledged to cooperate on reducing the consumption and production of hydro fluorocarbons (HFCs) by organizing a meeting of a bilateral task force on HFCs in advance of the next meeting of the Montreal Protocol, taking place in November 2014, and subsequently cooperating on next steps to tackle the issue. The bilateral task force will discuss, among other issues, safety, cost and commercial access to new and alternative technologies. The premiers recognized the need to use the institutions and expertise of the Montreal Protocol to reduce HFCs, while reporting and accounting for quantities reduced under the UN Framework Convention on Climate Change (UNFCCC).

The US-India Strategic Dialogue is propelled by the dynamic momentum achieved in the US-India relationship over the last decade and is in pursuance of the global strategic partnership for a better world that Prime Minister Singh and President Obama reaffirmed during their meeting in November 2009 and April 2010 in Washington. Secretary Clinton and Minister Krishna pledged to deepen people-to-people, business-to-business, and government-to-government linkages between the world's oldest and largest democracies, for the mutual benefit of both countries and for the promotion of global peace, stability, economic growth and prosperity. Both recalled that the India-US partnership rests on the firm foundation of common ideals as well as security and economic interests. The guiding principles upon which both nations were founded-democracy, mutual respect, individual liberty, rule of law, and an appreciation for the strength we derive from being pluralistic societies—make the India-US bond strong, resilient, and uniquely important for building a peaceful, prosperous, inclusive, and sustainable world.

They pledged that as strategic partners, India and the United States would continue to consult each other closely on regional and global developments, and remain sensitive to each other's interests. They confirmed that global institutions of governance should reflect contemporary realities and enhance effectiveness, in order to meet the challenges of new century. Minister Krishna and Secretary Clinton discussed progress on energy and climate change issues since the signing of the US-India Memorandum of Understanding on Energy Security, Clean Energy, and Climate Change. They stressed that the Clean Energy and Climate Change Initiative they finalized last year would contribute to achieving sustainable development and a clean energy future for the people of both countries. The goal of establishing a Joint Clean Energy Research Centre and accelerating collaborative efforts to deploy clean energy technologies in both countries remains a priority. They reiterated that the "Partnership to Advance Clean Energy," to be launched by both countries to take full advantage of multi-agency coordination and engagement with the private sector, constitutes an important step forward in developing and deploying clean energy.

They also welcomed the launch of the India-US Climate Dialogue, which seeks to enhance bilateral US-India cooperation with a view to achieving a successful outcome of the ongoing multilateral climate change negotiations. They reiterated the importance of the Climate Dialogue in focusing and coordinating both countries' priorities in dealing with climate change consistent with their national policies and strategies. In this regard, both sides recognized that further cooperation in knowledge on climate science can enhance our shared understanding of the planet's climate. The two sides reaffirmed their commitment to cooperate in the area of unconventional gas resources, as outlined in the Memorandum of Understanding to Enhance Cooperation on Energy Security, Energy Efficiency, Clean Energy and Climate Change. Climate change is one of the most important global environmental challenges facing humanity with implications for food production, natural ecosystems, freshwater supply, health, etc. According to the latest scientific assessment, the earth's climate system has demonstrably changed on both global and regional scales since the pre-industrial era. The Intergovernmental Panel on Climate Change (IPCC) projects that the global mean temperature may increase between 1.4 and 5.8°C by 2100. This unprecedented increase is expected to have severe impacts on the global hydrological system, ecosystems, sea level, crop production and related processes. The impact would be particularly severe in the tropical areas, which mainly consist of developing countries, including India.

The impact of climate variability and change, climate policy responses, and associated socio-economic development will affect the ability of countries to achieve sustainable development goals. The pursuit of these goals will in turn affect the opportunities for, and success of, climate policies. In particular, the socio-economic and technological characteristics of different development paths will strongly affect emissions, the rate and magnitude of climate change, climate change impacts, the capability to adapt, and the capacity to mitigate. For India and other developing countries, climate change is a potential threat to sustainable development. Through a multitude of government, private, and individual initiatives, the country is making serious efforts to conserve energy, harness renewables, protect forests, and enhance understanding about climate change. By adopting this climate-friendly path to development, India is contributing effectively to global efforts for protecting our planet.[21] The challenge of climate change is a global one and, fittingly, the global community is addressing this challenge through a multilateral and cooperative approach. The UNFCCC (United Nations Framework Convention on Climate Change) was adopted at the Rio Earth summit on 9 May 1992. The Government of India has demonstrated its deep and continuing commitment to global efforts to address climate change by acceding to the Kyoto Protocol to the UNFCCC, and hosting COP-8, the eighth session of the Conference of Parties to the UNFCCC.[22]

Climate Change is a matter of grave concern for developing countries like India, which are highly vulnerable to its potential impacts. The IPCC (Intergovernmental Panel on Climate Change) points out that developing economy rely more heavily on climate-sensitive sectors like agriculture, which operate close to environmental and climate tolerance levels. Few developing countries have the necessary financial, technical, and institutional capacities for efficient adaptation to climate impacts (IPCC 2001).[23] Changes in temperature and precipitation may adversely affect freshwater availability and quality in many areas. Climate Change could also aggravate problems of biodiversity loss and desertification. Climate change disproportionately impacts the poorest in society, exacerbating inequities in access to adequate food, clean water, health, etc. Thus, the issue of climate is closely linked to other environmental issues, and to the challenge of sustainable development itself.[24]

Although the countries of the development of the developing world are more vulnerable to climate change, their contribution to the greenhouse problem has been much smaller than that of developed countries. Historically, developed countries have been responsible for more than 60 per cent of GHGs (greenhouse gases) added in the last 100 years (WRI 2001). This is recognized in the UNFCCC, which follows the principles of 'common but differentiated responsibilities' and 'respective capabilities' in addressing its ultimate objective of stabilizing atmospheric GHG concentrations. It notes that the largest share of historical and current global emissions of GHGs has originated in developed countries, and enjoins upon them to take the lead in combating climate change.[25]

6.5 India-US Perceptions on Climate Change

This chapter presents India's perceptions on the problem of climate change and sustainable development; the kind of negotiating positions that follow from these perceptions; the politics India has undertaken so far and finally India's possibilities for action that can help contain the threat of climate change. The US politics surrounding the climate-change issue represent a tug-of-war between those who fear potential economic damage from restrictions on fossil-fuel combustion and those who fear for the health of the planet and the longer-term economic consequences resulting from climate change. Policies have swung from one pole to another with each new US administration. In March 2001, US President George W. Bush announced that the US administration would not support ratification of the Kyoto Protocol and on 11 June, 2001, he called the treaty "fatally flawed." "No one can say for certainty what constitutes a dangerous level of warming, and therefore what level must be avoided." He also attacked the agreement for not including developing countries, like China, the world's second-largest source of greenhouse gases. Thus, Protocol participants were left with two options: (1) adjust the Protocol to meet US objections, that is, force developing nations to also reduce emissions or (2) ratify the Protocol without the United States. Some European groups point to the huge investment in intellectual and research efforts that have gone into achieving the Kyoto Protocol. Renegotiating a new agreement would not necessarily improve the agreement and would delay implementation for years.[26] An associated challenge that India faces is serious energy shortage. Despite its conventional and unconventional resource potential, 57 per cent of the population lacks access to electricity (IEA 2002a). Access to sustainable energy is imperative to enhance economic opportunities and improve quality of life. Despite its low share in atmospheric GHG concentrations and its overriding development priorities, India is undertaking numerous initiatives that contribute significantly to international efforts for atmospheric protection, thus putting the country on the path of climate-friendly development.[27] The threat of climate change that led to the Framework Convention on Climate change (FCCC) at Rio, is perceived differently by different countries.

This fact has delayed any effective international agreement on how to deal with the problem. In the case of the Montreal Protocol covering ozone-depleting substances, there was a wide consensus and effective action was mobilised quickly. Thus, an understanding of perceptions and positions of different countries makes it easier to explore possibilities of effective action.

6.6 Climate Change Discourses in India

India is one of the world's most vulnerable countries to climate change. According to Thaker J and Leiserowitz A, the developing countries like India are under international pressure to sign a legally binding emissions treaty to avert catastrophic climatic change.[28] Developing countries, however, have argued that any international agreement should be based on historic and per capita carbon emissions, with developed countries responsible for reducing their emissions first and funding mitigation and adaptation in other countries. India, however, is now the world's third largest greenhouse gas (GHG) emitter, having tripled its carbon dioxide emissions from fuel combustion alone between 1990 and 2011 and its emissions are predicted to increase by almost 2.5 times between 2008 and 2035.[29] Climate change emerged as an issue for Indian policy-makers in the late 1980s, a decade which included the formation of the IPCC in 1988, at a time when the Indian government was beginning to transition to a market oriented economy to accelerate growth and reduce poverty. Several scholars argue that the initial Indian government positions on climate change were largely efforts to protect its domestic development and sovereignty in an increasingly unipolar world. In an ironic turn, the Indian government adopted principles articulated by Anil Agarwal and Sunita Narain—two long-standing critics of Indian government environmental policies —in their globally influential report, Global Warming in an Unequal World.[30] This report was written as a response to a directive by the then Indian Environment Minister to state governments to reduce agricultural and cattle-based emissions, based on a World Resources Institute report that had blamed India as one of the world's largest emitters, primarily because of methane emissions from agriculture and cattle stock.[31] Agarwal and Narain reframed the climate change issue as "environmental colonialism" making a distinction between "luxury" and "survival" emissions, and arguing that climate change has been primarily caused by the "historic" emissions of greenhouse gases by developed countries, not contemporary annual emissions.[32]

A framework to solve the problem, they argued, therefore needed to begin by allocating equal access to the atmosphere for all individuals in the world, the per-capita allocation principle. The Indian government drew heavily upon this report and in fact successfully inserted the terms "historical emissions" and "common but differentiated responsibilities" (CBDR) into the preamble of the UN Framework Convention on Climate Change. Moreover, there was a fear that global environmental concerns would now be used to limit and control developing countries' economic growth, through global financial institutions such as the World Bank and International Monetary Fund (IMF), controlled by the rich North. Others argued that climate change would divert the "attention of policy-makers, scientists and NGOs, often the first rate ones, from more pressing issues" especially, as the Director General of the Indian Meteorological Department suggested at the time, "in the absence of hard evidence".

Yet at the same time, India took a number of domestic measures and played an important role in international meetings on global environmental problems, including the formation of the UN Framework Convention on Climate Change (UNFCCC). Similarly, India played a major role at COP 1 in Berlin in 1995, submitting a proposal for a 20 per cent mission reduction by industrialized countries in the year 2000 compared to 1990 levels, a watered-down version of which ultimately became the 1997 Kyoto Protocol—to this date the only binding emissions reductions treaty. India was initially skeptical of market-based mechanisms such as the Clean Development Mechanism but due to increased lobbying by Indian business for access to international markets; it quickly established the national CDM project approval authority in December 2003 and is currently the second largest CDM host country in the world.

According to Dubash N.K., India occupies an intriguing dual position in global climate politics.[32] As a developing economy with a substantial poverty problem and low levels of historical and per capita emissions, India is both likely to face considerable impacts and has limited capacity to address the problem. As a large country and one that has transitioned into a rapidly growing economy, India gets called on to participate actively in addressing the global climate challenge. Indian climate politics has substantially been shaped around the first perspective, and increasingly, under international pressure, is being forced to grapple with the second.

6.6.1 India's Security Discourses and Climate Change

According to Brodie J. the term 'Security' has multiple and contested meanings that signifies the pre-cautions, agents, or organizations responsible for protection of individuals against a variety of harms ranging from criminal violence to espionage and sedition (i.e., security forces), or an assurance or guarantee, often in the form of a document indicating ownership.[33] Security also radiates a 'feel' when it is used to connote a highly subjective state of being, a certain sense of confidence, which accompanies a feeling of freedom from danger, anxiety, financial cares, or material privation. Brodie J also stated that security discourses are integral to nationalist difference making, as those identified as the threatening or foreign 'other' provides the mirror opposite of the national 'we'.[34] Critics of canonical security discourses stress that national security threats do not exist 'out there,' prior to or independent of their discursive construction and the security problems are constituted as such through specific discourses that lend urgency, legitimacy, and resources to their resolution. The national security discourse is focused on limited terms, i.e. foreign policy, military power, military capabilities the articulation of social security needs have been left to the socially concerned whose voice is not heard in the security arena.

In the Indian perspective the term security can be defined broadly or narrowly in both national and social context. Broader interpretations bring in the requirement of economic and social dimensions of a national security. When approached narrowly, it focuses on territorial integrity and sovereign power in the national security and the violation of political rights.[35] In modern India the concept of national security ought to be located in the interest of human welfare and national power. There is more to national security than territorial integrity and the preservation of state sovereignty. The improvement and development of human conditions is at least as important as achievement in the international stage. Micro-level well-being of the individual ought to be of as much concern as the macro-level stature of the nation. The security discourse in India is dominated by the aspirations of elite and the co-opted middle class. In India external and internal component of security is important. The external component has an economic and politico-military dimension. The internal component has a socio-economic and a politico enforcement dimension.[36]

It was in the 1970s that the environmental problems began to gain relevance in political discourse. The debate on environmental security gained momentum in the 1980s with the emergence of global environmental problems such as the depletion of stratospheric ozone or global warming. The term 'environmental security' gained importance into international debates when the World Commission on Environment and Development published 'Our common future' in 1987 and this marked the merits of this concept were considered on analytical and normative grounds. 'Environmental conflict discourse' is characterized by a relatively narrow view on the relationship between security and the environment and 'environmental security discourse' highlights the negative impacts of environmental degradation for human beings. While the environmental conflict discourse can be linked directly to traditional understandings of military security and the security of states, the environmental security discourse is much more closely linked to notions of "human security" in which the protection and insurance of human welfare is of central concern.[37] A 1994 United Nations Development Programme report outlines seven areas of human security.

(1) Economic security
(2) Food security
(3) Health security
(4) Environmental security
(5) Personal security
(6) Community security, and
(7) Political security

Detraz N. and Betsill M. also explored the implications of the April 2007 United Nations Security debate on the security dimensions of a changing climate for international climate change politics. Specifically, their analysis was focused on whether and how security concerns have been addressed in past international political debates on climate change by considering whether the Security Council debate, which emphasized the threat of climate-related conflict, reflected a discursive shift.[38] Thaker J and Leiserowitz A investigated the factors that are responsible for driving shifting Indian discourses of climate change by conducting and analysing 25 interviews of Indian climate policy elites, including scientists, energy policy experts, leading government officials, journalists, business leaders, and advocates, in addition to analysis of articles published in Economic and Political Weekly (a prominent Indian policy journal), and reports published by the government and other agencies.[39] Their analysis suggested that India's concerns about increasing energy access and security, along with newer concerns about vulnerability to climate change and the international leadership aspirations of the Indian government, along with emergence of new actors and institutions, has led to plurality of discourses, with potential implications for India's climate change policies.

India, however, is now the world's third largest greenhouse gas (GHG) emitter, having tripled its carbon dioxide emissions from fuel combustion alone between 1990 and 2011 and its emissions are predicted to increase by almost 2.5 times between 2008 and 2035 (IEA 2013). According to National Academies (2010) citing the rising national emissions of developing countries like India, developed countries like the United States have refused to sign a legally binding treaty to reduce greenhouse gas emissions. The climate change discourse has shifted character in a qualitative sense in the last few years, underscored by an increasing sense of urgency.[40] Risbey J reviewed the evidence to support claims that climate change can be viewed as 'catastrophic', 'rapid', 'urgent', 'irreversible', 'chaotic', and 'worse than previously thought' by using the method to assess whether the conventional understandings of these terms are broadly consistent or inconsistent with the science, or else ambiguous.[41] India's view of its own role and responsibility, and that of developing countries, has remained remarkably consistent through the years of its engagement with the issue. And today, as the international community is engaged in crafting a framework for future actions on climate change, India's strongly held and apparently rigid position is perceived as threatening the ability of the international community to arrive at a global deal.[42] The elements of India's foreign policy on climate change, including aspects of India's negotiating positions in the UN Framework Convention on climate change (FCCC) and Kyoto Protocol processes examines in particular India's interventions in the ongoing negotiations for future action on climate change. It also explores, in an initial and contingent fashion, the domestic roots of India's position.[43] A few caveats are in order. There is no singular conception of climate change as a phenomenon, or an articulation of it as an issue. There are differences across the spectrum of engagement on this issue-scientific, political, economic and rhetorical. Nations, epistemic communities, civil society organizations, business and industry, among others, differ in their perception, articulation and social construction of climate change. Some perceive it as a "threat" to economic advancement others as an "opportunity" for green growth. Some lay claim to authoritative and "objective" scientific knowledge, and seek a privileged role in policy-making. Others dispute such claims to knowledge and the value of experts over non-experts, and privilege cultural understandings of climate. Some perceive it as a global environmental problem, others, as a local and social problem as well. There are fundamental differences in the values that underlie both perceptions of the problem and its gravity, as well as possible solutions and burden-sharing arrangements. Developing and industrial countries, and communities within these countries, differ on questions of responsibility for and ways to address climate change.

There is neither a homogenous community of interests on climate change, nor is it a settled political space and here it has not presumed or suggested that there is. Let's proceed on the basis that climate change as an issue is unsettled and deeply contested. It seeks to locate India's official framing of the issue and it's positioning within the spectrum of variegated views and constructions of climate change.[44]

6.6.2 Elements of India's Foreign Policy on Climate Change

India is on a mission to develop economic growth, and with it, poverty eradication, energy security, and provision of universal access to energy is a central, enduring, and obsessive preoccupation of the Indian government. And, justifiably so, India is placed 128th on the Human Development Index 10, 34.3 percent of its population lives on less than US$ 1 a day, and an estimated 44 percent does not have access to electricity. India's developmental mission, as framed, however, may well leave large carbon footprints, and ultimately weaken its ability to develop. Its energy use, now at a low per capita emissions rate of 1.2 metric tons annually, and a cumulative share of 4.6 percent, will change significantly in the near future. If India's current growth rate continues energy demand will more than double by 2020. In addition, if India's targets on poverty, unemployment, and literacy in its 11th five year plan—some more ambitious than the Millennium Development Goals (MDGs)—are to be met, and energy provided to the 498 million Indians without access to electricity, it will lead to much greater energy use. India will soon be a significant contributor to climate change. Indeed, India is predicted by some estimates to become the third largest emitter by 2015, and with US, EU, China, and Russia, to account for two-thirds of the global greenhouse gases. Unsustainable Consumption Patterns of the rich industrialised nations are responsible for the threat of Climate change. Only 25 per cent of the global population lives in these countries, but they emit more than 70 per cent of the total global CO emissions and consume 75 to 80 per cent of many of the other resources of the world (Parikh et al 1991). In per capita terms, the disparities are also large: an Indian citizen emits less than 0.25 tonnes of Carbon per year whereas a citizen of the USA, for example, emits more than 5.5 tonnes. Parikh et al., emphasize the need for an equitable and efficient solution to climate change and suggest that efficiency can be obtained through an off tradable emission quotas and equity through equal allocation of global environmental space to all human beings. These findings were well received by developing countries and are echoed in the UNFCCC, which recognised the rights of developing countries to economic development and also the "common but differentiated responsibilities" of different countries.

6.6.3 Methane Emissions and Subsistence Emissions

Another theme of Indian analysts has been the lack of reliability of GHG emission estimates, particularly of methane. According to initial estimates, large emissions of methane from paddy fields were ascribed to developing countries. However, the empirical basis of these estimates was questioned; subsequently experimental measurements by Indian researchers showed these doubts to be well-founded (Mitra, 1992, 1996). Moreover, Agarwal and Narain (1991) argue that emissions by poor who live on the margin of subsistence should be considered a basic human right and should not be counted when ascribing responsibilities for emission reduction.

6.6.4 Why should India be concerned about Climate Change?

The Earth's environment acquired salience as an issue of international concern with the convening of the UN Conference on the Human Environment at Stockholm in 1972. It brought environmental issues into the ambit of international diplomacy and led to gradually enhanced global environmental cooperation. Addressing the Conference, Indira Gandhi drew attention to the link between environment and development, highlighting that "the environment cannot be improved in conditions of poverty." The Conference led to the establishment of the UN Environment Programme (UNEP), which in 1982 convened in Nairobi a 'UNEP Session of a Special Character: Ten Years after Stockholm'. It recognized that most global environmental challenges remained inadequately addressed and environmental threats had grown, including from acid rain, air, soil and water pollution, desertification and deforestation, ozone layer depletion. Early signs of the disruptive potential of climate change were also noted.

As fallout of the Special Session, the UN Secretary-General appointed the World Commission on Environment and Development in 1983. Its Report, "Our Common Future", was released in 1987 and sought to balance human and environmental well-being and reconcile economic development with environmental protection. It promoted the concept of "sustainable development", which it defined as "development that meets the needs of the present without compromising the ability of future generations to meet their own needs". This definition viewed environment and development through the prism of "needs", in particular those of the world's poor, and highlighted the notion of intergenerational equity.

By the time the UN Conference on Environment and Development was convened at Rio de Janeiro in 1992, one could not talk of the environment without referring in the same breath to development. That Summit adopted the Rio Declaration on Environment and Development and Agenda 21, which reinforced the importance of finding ways to generate economic growth and address the overriding priority of development for developing countries and doing so without hurting the environment. Two important legal instruments, the UN Framework

Convention on Climate Change (UNFCCC) and the Convention on Biological Diversity, were also opened for signature at Rio.

In 2019, in Chennai, locals were praying for some rain; in Mumbai, people were reeling under a deluge. Long ago, these extreme disparities may have been solely blamed on nature's vagaries, but now science has established that human-induced climate change is playing a major role. Climate change, caused by emissions from industries and other human activity, is making the world warmer, disrupting rainfall patterns and increasing the frequency of extreme weather events. No country is immune to these forces, but India is particularly vulnerable. In 2018-19, as many as 2,400 Indians lost their lives to extreme weather events such as floods and cyclones, according to the environment ministry. The India Meteorological Department (IMD) says these events are increasing in both frequency and intensity Extreme events may be the most tangible and immediate impact of climate change, but another more long-term and equally dangerous effect is rising temperatures.

In India, according to IMD data released by the statistics ministry, average temperatures have increased by 0.6°C between 1901-10 and 2009-18. At an annual level, this may seem trivial, but projections deeper into the future paint a more alarming picture. For instance, the World Bank estimates that, if climate change continues unhindered, then average temperatures in India could reach as high as 29.1°C by the end of the century (up from 25.1°C currently). Few countries are likely to suffer from climate change to this extent. According to the Global Climate Risk Index released by Germany-based think tank, German watch, India is the 14th most climate change-affected country in the world. This vulnerability, though, is not India's own doing. In terms of global greenhouse gas emissions, India's share remains significantly lower than those of both the US and China. In many ways, India is paying for the excesses of the developed world.

6.6.5 The Challenge of Climate Change in India

The harsh consequences of climate change will intensify challenges to development, especially in India, Bangladesh, Maldives, Sri Lanka and other South Asian countries. Natural hazards and vulnerabilities arising from economic, social and environmental circumstances already make

India one of the more disaster-prone countries of the world. Climate change related risks will increasingly affect the Indian subcontinent, including via sea level rise, cyclonic activity and changes in temperature and precipitation patterns. Rising sea levels would submerge low-lying islands and coastal lands and contaminate coastal freshwater reserves. Melting Himalayan glaciers would reduce downstream water supply in many of India's important rivers in the dry season, impacting millions. A warmer atmosphere will spread tropical diseases and pests to new areas. Moreover, India's urban population, projected to rise from 377 million in 2011 to 745 million in 2041, is already seriously challenging services and infrastructure in stressed Indian cities. Indeed, most Asian cities with over 1 million inhabitants are already exposed to water related hazards. Climate change will increase the risks of death, injury and ill-health and disrupt livelihoods in low-lying coastal zones due to cyclones and coastal and inland flooding, storm surges and sea-level rise. Increased river, coastal and urban floods could cause considerable loss of life and widespread damage to property, infrastructure and settlements. Extreme rainfall and flooding will cause deaths, illnesses and mass displacement. The impact of climate change on tropical cyclones would vary by region, but projections across models agree on more frequent and heavy rainfall days and that an increase in extreme rainfall events related to monsoons is very likely in South Asia. Alteration in the Indian monsoon pattern can also be expected, with the volume of rainfall during the monsoon season set to rise. However, not only would there be more water, the way in which it is delivered would also change. Moreover, while the number of rainy days is expected to be less, the intensity of rainfall would be more. Record high temperatures are already occurring more often. Temperature trends show that the frequency of hot days in India is likely to increase further in the future. The severe flooding in Mumbai in 2005 suggests interaction between climate change and other stressors. The flash floods in Kedarnath in 2013 and Srinagar in 2014, Uttarakhand fires in 2016 and Vadodara flood in 2019 also bring out the need for considerably better disaster preparedness. The floods in Assam are a calamity that annually takes lives, harms livelihoods and damages infrastructure. In many instances, restoring forests, floodplains and wetlands are crucial to address annual flooding by rivers. Floods due to glacial lake outbursts are another emerging threat in parts of the Himalayas.

6.6.6 Impact on Human Health

Extreme weather events often also collapse electricity, water supply, and health and emergency services. Moreover, increased heat-related mortality and heat stroke due to rising and extreme temperatures could undermine the modest progress made in recent decades by South Asian countries in tackling disease, malnutrition and early deaths. Disease causing pathogens and parasites will multiply faster at higher temperatures, escalating the incidence of many tropical diseases. Japanese encephalitis and dengue fever outbreaks are also associated in India with high temperature and rainfall. The presence of Malaria in India has been linked to rainfall patterns. Studies show an association between higher temperatures, heavy rainfall and outbreaks of diarrhoea and cholera. Urban populations will also suffer greater heat stress. In urban areas, where child mortality is already high, extreme temperatures will lead to more deaths. Mental disorders and post-traumatic stress syndrome could also be seen in disaster-struck areas. Contaminated urban flood waters will increase exposure to disease and many toxic compounds.

6.6.7 Impact on Food Production and Livelihoods

Climate change will seriously impact global food production as drought, increased unpredictability of precipitation, and rising temperatures would reduce global crops yields, while warming and acidification of the oceans would affect marine wildlife and fisheries. Most of the food-insecure are in South Asia, where currently over 400 million poor and undernourished people live. Climate change will disproportionately hurt many of them. More erratic rainfall in parts of India could lower rice yields and lead to higher food prices and living costs, while increased drought related water and food shortages linked to rising and extreme temperatures may increase malnutrition and worsen rural poverty. Over 55 per cent of Indian rural households depend on agriculture for a living and, with fisheries and forestry, it is amongst the larger contributors to India's GDP. Many Indians who earn livelihoods in coastal regions will also be affected.

6.6.8 Heat stress could lead to substantial food yield and production declines

Climate change will also negatively impact livelihoods through its effects on ecosystems, some of which are highly vulnerable. The geographic range, seasonal activities and migration patterns of many land, freshwater and marine species have shifted in response to ongoing climate change. Abundance of species has changed. Permafrost degradation has been reported for parts of the Tibetan Plateau. Earlier greening has been observed in Himalayan forests and could increase their vulnerability to wildfires. Coral reefs of Lakshadweep and Andamans are bleaching due to higher sea temperatures.

6.6.9 Responding through Adaptation and Mitigation

The response to climate change has to be through both adaptation and mitigation. We must adapt our societies to prepare for some climate change risks. Adaptation delivers many immediate and future benefits, but has its limits. Improving ecosystem resilience and helping local communities are low-regret path to successful adaptation, especially in India. Opportunities for implementing low-regrets adaptation measures emerge, for example, by reducing water use via drip irrigation and water recycling; enhancing water harvesting and water storage; planting drought-resistant crop varieties; enabling cities to cope with increasingly frequent extreme weather events; building new infrastructure to protect coastal cities from sea-level rise; using more porous materials for berms/ footpaths in urban settings; improving land governance and ensuring security of land tenure; boosting disaster-relief preparedness; building social protection systems and safety nets; raising skill levels for better absorption of new technologies; and involving stakeholders from the outset in planning processes. Mitigation efforts must be intensified given the early benefits they can deliver. The world needs to significantly reduce GHG emissions to limit the magnitude of climate change. There are several ways of doing so, including by switching from conventional fossil fuels (oil, coal and natural gas) to low or renewable carbon energy sources such as wind, solar, ocean, geothermal, hydroelectric, and nuclear energy, as also by energy conservation and more efficient energy usage (e.g. via more fuel efficient vehicles, spread of Metro mass transportation to more

Indian cities, and use of less energy consuming LED bulbs and home appliances) and reducing energy wastage, as well as by more effective urban planning (so as to lessen vehicle usage and emissions) and better building design. Mitigation can also be achieved by increasing the capacity of carbon sinks, e.g. through reforestation and afforestation, while removal and burial of carbon dioxide can also be explored.

India has one of the lowest rates of energy intensity of GDP growth. India's historic and current levels of per capita GHG emissions remain the lowest amongst the G20, which constitute the major world economies, even though in volume terms it is now the third-largest GHG emitter in the world, after China and USA. India continues to face massive development challenges to tackle which it requires a high level of sustained economic growth. It was not a major GHG emitter in 1992, but is today in the spotlight for its rising GHG emissions. India's per capita GHG emissions remain a fraction of that of all major emitters and it does not want to pursue the environmentally harmful development unwittingly followed by developed countries in their process of industrialization. However, India cannot finance the huge investments needed to enhance its capacity to adapt to climate change unless its economy grows at a fast tempo. India seeks to meet the climate change challenge by expanding the use of low carbon and renewable technologies and improving energy efficiency of buildings, factories, appliances, etc. In the long run some low-carbon development options may be less costly for India and could offer new economic opportunities. Inclusive growth too is integral to an effective climate change policy for India and studies in India show that low carbon growth pathways are consistent with inclusive growth.

6.7 India: Paris Climate Deal

The year 2015 is critical for addressing sustainable development and climate change. In mid-July, the 3rd International Financing for Development Conference was held in Addis Ababa but has regrettably not come out with fresh commitments by the developed world on climate finance. The Paris Agreement is an agreement within the United Nations Framework Convention on Climate Change (UNFCCC), dealing with greenhouse-gas-emissions mitigation, adaptation, and finance, signed in 2016. As of November 2019, all UNFCCC members have signed the agreement, 187 have become party to it, and the only significant emitters which are not parties are Iran and Turkey. India is taking actions to fulfil its pledge made at the Copenhagen Summit in 2009 to bring down the emission intensity of its GDP by 20-25 per cent by 2020 over 2005 levels. Its response to climate change must be guided by its national plans, programmes and priorities, not by the push and pull of international negotiations. It should encompass enhanced adaptation measure and also mitigation measures that restrict GHG emissions without compromising its development compulsions, more so if incremental costs and technologies for its mitigation efforts are borne by the Green Climate Fund or the developed world. India can achieve a similar level of well-being as the developed world without going down a path of reckless and wasteful consumption. It has in the past indicated that its per capita emissions would never exceed those of the developed countries, including their historical emissions. It is also prepared to share its technologies with others, as seen from its offer of free-of-cost remote sensing satellite data to other SAARC countries.

"The Paris agreement is undoubtedly a turning point in the history of common human endeavour, capturing the combined political, economic and social will of governments, cities, regions and businesses and investors to overcome the existential threat of unchecked climate change." This chapter has summarized the climate policy issues currently facing India at a time when the global climate debate has stalled around the lack of progress in the Kyoto Protocol. It is particularly important to have this debate in India when more than a billion people are poised to enter a new phase of higher economic growth. It is argued that there are policies such as the McKibbin Wilcoxen Blueprint, which could be implemented in India as part of a broader strategy of action that both prices future carbon emissions and encourages economic development. The development of institutions to manage risk as well as the clear commitment to taking effective action against future carbon emissions has the potential to be an attractive option for a country like India. If it was shown to be successful in both stimulating foreign investment in energy development and reducing the trend of greenhouse emissions through market based incentives based on the clear establishment of property rights, the demonstration effect across the developing world would be powerful. It would certainly remove the complaint by countries like the United States and Australia against taking action because of lack of binding commitments by developing countries. That alone would reduce greenhouse gas emissions significantly in future decades. The alternative strategy for India is to wait for a resolution of the stalemate over the Kyoto protocol and for large sums of financial assistance to accompany the transfer of energy technology from the industrial economies though some other Kyoto like endeavour.

In COP21 in Paris in 2015, India accepted 1.5°C as a target limit for the increase in the global average temperature and launched the global renewable energy initiative, International Solar Alliance (ISA). It has also since announced an ambitious domestic renewable energy programme. At the World Economic Forum meeting in Davos in February 2018, Indian Prime Minister called climate change one of the biggest global threats and called for action to prevent it. Indian Prime Minister's choice to pay attention to climate change at a forum traditionally focused on economics and foreign policy is one example of how the significance of climate change is being recognised to the extent that its consequences extend to global politics and security.

Indian Prime Minister spoke at the climate action summit organized by the UN Secretary-General in September-October 2019. He highlighted what India has done to address climate change issues and his expectations from the international community. Prime Minister invited UN member states to join India's Initiative on Coalition for Disaster Resilient Infrastructure. PM inaugurated a 50 KW 'Gandhi Solar Park' on the roof of the UN Headquarters. India gifted the $1 million solar park, which has one solar panel for each of the 193 UN Member states.

6.7.1 How US's Exit from Paris Deal affected India

The India-US relationship can be characterised by the good, the bad, and the undecided. On the plus side, despite their differences, the two countries found common ground on three international agreements on climate change (in Paris, under the Montreal Protocol to phase out hydrofluorocarbons, and measures to curb emissions under the International Civil Aviation Organisation). Bilaterally, they launched numerous programmes on energy access, and established joint funds to push energy R&D. The negatives include contentious issues such as differentiation between countries' responsibilities and capabilities under the Paris Agreement, how financial contributions would be counted, and several trade disputes on clean energy. Further, it is unclear how the US withdrawal would immediately impact bilateral energy programmes, joint investments, and investor confidence.

First, energy prices will matter for any long-term contracts. GAIL is contracted to purchase 5.8 million metric tonnes per annum (MMTPA) of liquefied natural gas (LNG) from two US terminals. But the landed price of the gas will be higher than from sources in West Asia or Africa. That said, by 2020, India will have the capacity to import 50 MMTPA of gas. If the price were right, in principle there would be a lot of scope for additional gas imports (in fact, up to 88% of projected net LNG exports from the United States). Equally, India signalled its continued ambitions on nuclear energy by sanctioning ten indigenous reactors (a total of 7000 MW). As cleaner energy sources, both gas and nuclear could offer benefits for both countries, but fuel exporters and technology providers would have to cut their costs.

Second, energy efficiency offers huge technological and commercial opportunities. If India's high economic growth rates have to be maintained within a shrinking carbon space, energy efficiency will matter in the industrial, transport and buildings sectors. Currently, only large industrial units are covered under countrywide energy efficiency schemes. However, micro, small and medium enterprises (numbering over 3 million and located in 200 industrial clusters) have a non-trivial share of industrial energy use (about 14%-15%). India's space cooling needs will grow five-fold in two decades, again offering scope for Indian and US firms to collaborate on technologies. Growing transportation demand will mean an expanding market for batteries for electric vehicles, fuel cells and third-generation biofuels. None of these market opportunities require the Paris Agreement per se.

Third, energy security matters for both countries. India and the United States need to raise the dialogue on maritime security cooperation in the Indian Ocean region, to ensure security of energy supply routes. Another emerging area of concern is the security of the electricity grid against cyber-attacks. There is already a joint programme on transforming India's grid. Grid stability and cyber-security could offer a sustained basis for cooperation, technology exchange and commercial investment.

Fourth, the uncertainty in US federal policies towards renewable energy has made India's renewable energy market relatively even more attractive, not just for its size (175 GW by 2022) but for the investment opportunity that credible, long-term policies offer. India saw nearly $10 billion invested, both in 2015 and in 2016, in renewable energy projects. Last year, $1.9 billion of green bonds were issued. India's solar targets, alone, need $100 billion of debt. These are huge investment opportunities for US-based firms, as well as manufacturers. The past decade had many instances of transformational deals (bilateral and multilateral) for India and the United States. President Trump's announcement on the Paris agreement is highly unfortunate, but does not close the door completely.

6.8 THE US PERCEPTION

The issue of global warming has always been a great concern for the United States of America. Through the years, the US has had one of the world's highest per capita emissions for greenhouse emissions. Since 1895, the US has maintained a national record of temperatures and 2016 was recorded as the warmest year in the US.[45] In 2009, the National Oceanic and Atmospheric Administration (NOAA) issued a statement which highlighted the impacts of global warming. According to the statement issued, "lake and river ice is melting earlier in the spring and forming later in the fall. Plants are blooming earlier in the spring, mountain glaciers are melting, coastal temperatures are rising and a multitude of species of birds, fish, mammal and plants are extending their ranges northward and in mountainous regions, upward as well."[46] Due to the rise in greenhouse emissions and its consequential effect on climate change, the US deems climate change as a national security risk. Therefore, in order to tackle climate change and safeguard the US's security interests, President Obama introduced the Copenhagen Accord, which would lay the foundation for international action on climate change. In order for the Copenhagen Accord to be globally accepted, President Obama ensured that the US, the EU, India, China and Japan were signatories to the said accord. Having the aforementioned signatories ensured that those responsible for 80 per cent of the world's global warming pollution from deforestation and energy, would work together to reduce not only their individual greenhouse emissions, but also to work together to combat global climate change.[47]

6.8.1 The Cancun Agreement

While the US did not ratify the Kyoto Protocol, the Cancun Agreement can be considered more of a success. The Cancun Agreement has signatory nations commit to the creation of a Green Climate Fund. In addition to this, the Cancun Agreement also stressed countries to uphold the Copenhagen Accord to reduce their greenhouse emissions. One of the key takeaways for the US during the Cancun summit was to push for a more robust and transparent policy towards developing countries. In addition to this, the Cancun Agreement also included transparency and accountability requirements for all countries. All signatories to the Cancun Agreement were obliged to provide a timeline for implementing actions to reduce their individual greenhouse emissions. In the Cancun Agreement, the US pledged to reduce its greenhouse emissions by 17 per cent below its 2005 levels.[48] One of the key successes for the US with this Agreement, was that the US succeeded in developing mechanisms to hopefully support the needs of developing countries, help in innovation, research and development, and the innovation of new technologies, and help track their emissions in the process.[49]

6.8.2 The Durban Agreement

The Durban Agreement was considered as a major success, as it brought on board the world's three largest global emitters, the US, India and China. Together, these three make up half the global emitters.[50] Despite signing the Kyoto Agreement, the US did not ratify it. However, with the Durban Agreement, the US agreed to legally binding cuts in its emissions. The Durban Agreement will come into effect only in 2020.[51] Moreover, the Durban Agreement also ensures that the Clean Development Mechanism will continue to work in the future.[52] In addition to this, the Durban Agreement also laid the foundation for the establishment of the Green Climate Fund, which would help disperse US$ 100 billion a year to poor countries to help adapt to the impact of climate change.[53] The Durban Agreement asserted that climate change should be tackled through International Law and not through national law and voluntarism.[54]

6.8.3 The Paris Agreement

The Paris Agreement is one of the most important agreements with regard to climate change. From the 197 member countries of the UN, 186 states and the EU, representing more than 87 per cent of global greenhouse gas emissions, have ratified or acceded to the Agreement, including China, the United States and India, the countries with three of the four largest greenhouse gas emissions of the UNFCCC members . Enforced on November 4, 2016, the Paris Agreement ensures that those who have ratified the agreement agree to maintain a global temperature rise below 2°C throughout this century.[55] One of the most unique features of the Paris Agreement is that the agreement focused more on nationally determined contributions.[56] The Paris Agreement ensures that each country is responsible for controlling its own emissions, depending on it industrial and economic capabilities. In addition to this, each country will set its own targets to meet, rather than have global targets enforced on them. The only target to be met by all the ratified members is to maintain a global temperature rise below 2°C. Furthermore, the document clearly states, "Developed countries should continue taking the lead by undertaking economy-wide absolute emission reduction targets. Developing countries should continue enhancing their mitigation efforts, and are encouraged to move over time towards economy-wide emission reduction or limitation targets in the light of different national circumstances."[57] This statement guides countries like India to industrialise and simultaneously gradually shift from using fossil fuels to using renewable energy sources. In addition to this, statement helps every developing country to industrialise itself to the point where a blanket agreement can be enforced globally. In order to ensure that every country meets its required target of curbing its emissions, the Paris Agreement reinforced a global stock care every five years, to ensure that everyone is on track to reducing their emissions. The first global stock care will be undertaken in 2023.[58] With national goals being set out in the Paris Agreement, the agreement worked in favour of the US. The task of monitoring and controlling emissions was then entrusted to the EPA under the Obama administration.

6.8.4 Withdrawal from the Agreement

On June 1, 2017, United States President Donald Trump announced that the US would cease all participation in the 2015 Paris Agreement on climate change mitigation. Trump stated that "The Paris accord will undermine (the US) economy," and "puts (the US) at a permanent disadvantage." During the presidential campaign, Trump had pledged to withdraw from the pact, saying a withdrawal would help American businesses and workers. Trump stated that the withdrawal would be in accordance with his America First policy.

In accordance with Article 28 of the Paris Agreement, a country cannot give notice of withdrawal from the agreement before three years of its start date in the relevant country, which was on November 4, 2016, in the case of the United States. On November 4, 2019, the administration gave a formal notice of intention to withdraw, which takes 12 months to take effect. So, the earliest possible effective withdrawal date by the United States cannot be before November 4, 2020, four years after the Agreement came into effect in the United States and one day after the 2020 US presidential election. The White House later clarified that the US will abide by the four-year exit process. Until the withdrawal takes effect, the United States may be obligated to maintain its commitments under the Agreement, such as the requirement to continue reporting its emissions to the United Nations. While celebrated by some members of the Republican Party, international reactions to the withdrawal were overwhelmingly negative from across the political spectrum, and the decision received substantial criticism from religious organizations, businesses, political leaders of all parties, environmentalists, and scientists and citizens from the United States and internationally. Following Trump's announcement, the governors of several US states formed the United States Climate Alliance to continue to advance the objectives of the Paris Agreement at the state level despite the federal withdrawal. As of July 1, 2019, 24 states and Puerto Rico had joined the alliance, and similar commitments had also been expressed by other state governors, mayors, and businesses.

6.8.5 The United States Environmental Protection Agency

The EPA is tasked with protecting human health and the environment, national efforts to help reduce environmental risks based on scientific facts and information and environmental protection.[59] To protect Americans, the EPA implements laws as soon as they are enacted by Congress[60] and also sets national standards that states are required to meet through their own regulations.[61] The EPA was responsible for implementing the Clean Air Act. This federal law regulated air emissions from stationary and mobile sources and authorised the EPA to establish National Ambient Air Quality Standards (NAAQS) to protect public health and welfare.[62] Through the years, the Act has been amended over the years to set new goals and dates for achieving NAAQS, as most states in the country had failed to meet the deadlines.[63] In addition to this, the Clean Air Act regulates emissions from mobile and stationary sources. When it comes to stationary sources, the Clean Air Act gives the EPA the power and the authority to take action to eliminate or minimise emissions from coal-fired power, acid, glass and cement plants and petroleum refineries.[64] The emissions from these plants and refineries can contribute to respiratory illness, heart disease, the formation of acid rain and reduced visibility to name a few.[65] While the EPA regularly monitors emissions from stationary sources, it also monitors emissions from mobile sources. The EPA ensures that motor vehicle engines and off-road vehicles and engines meet the Clean Air Act requirements. While importing a vehicle to the US, the said vehicle must have an EPA issued certificate to conform to the Clean Air Act before its import.[66] The EPA thus ensures that each vehicle entering the US conforms to all its emissions requirements. Despite the efforts made by the EPA to ensure reduced emissions, there has been a steady flow of illegally imported and uncertified motorcycles and vehicles in the US since 2008.[67] According to EPA estimates, illegally imported cars and motorcycles are responsible for 30 per cent higher emissions above the allowable standards.[68]

In addition to these measures, the Clean Air Act also makes it illegal to manufacture, sell or install a part in a motor vehicle that would bypass or defeat the purpose of an emission control device.[69] Moreover, the Act also prohibits people from tampering with an emission device. If an emission device has been tampered with, the Clean Air Act gives the EPA the right to persecute the person responsible for tampering with the device. In order to prevent tampering, the EPA works with companies to ensure that their devices are tamper proof. The EPA also regulates fuel used in vehicles. For this, there are specific EPA guidelines to regulate fuel and additives sold in the US. Moreover, the EPA also ensures that there is some amount of renewable fuel in transportation fuel to reduce greenhouse emissions. Refiners and importers of fossil fuels need to show compliance with the EPA regulations of using renewable fuels.[70] In addition to the Clean Air Act, the US Congress has also implemented a Clean Water Act that establishes the basic structure of regulating water pollutants. Under this Act, the EPA has implemented pollution control programs, such as setting wastewater standards for the industry.[71] The Clean Water Act has also made it illegal to discharge any pollutants in the water at their sources. The Act helps the EPA monitor water emissions through the National Pollutant Discharge Elimination System (NPDES).[72] Through the NPDES, the EPA reviews the reports sent by companies to show their compliance, and it also conducts an on-site review to ensure that companies show their compliance to the norms of the NPDES. This compliance takes place more on a state level than just a national level.[73] Furthermore, the Clean Water Act NPDES Compliance Monitoring Strategy provides implementation guidelines to the EPA and the authorised states by constantly describing the EPA's frequency goals.[74] Section 404 of the Clean Water Act helps regulate the placement of dredged or fill material into wetlands, lakes, streams, rivers, etc. The main goal of Section 404 is to avoid and minimize losses to wetlands and other waters and to compensate for unavoidable loss through mitigation and restoration.[75]

The US Army Corps of Engineers works hand in hand with the EPA to implement Section 404. One of the most salient featured of the Clean Water Act implemented by the EPA, is that it prohibits the discharge of oil and other hazardous substances in US waters as well as along its shoreline.[76] In order to transport oil, the EPA Oil Pollution Prevention regulations require owners and operators to implement plans to avoid oil spills.[77]

Thus, through the Clean Air and Clean Water Acts, the US Congress has empowered the EPA to ensure the air and water safety in the US. Keeping Americans healthy and reducing emissions, is the bedrock of the EPA. Under the Obama Administration, the EPA flourished and enacted policies to help gradually reduce emissions from manufacturing plants and other sources; under the Trump administration, things have been a little different. He has questioned if climate change is real and has indicated that he will focus his efforts on other causes as president. Trump has also expressed that efforts to curb fossil fuel industries hurt the United States' global competitiveness. He pledged to roll back regulations placed on the oil and gas industry by the EPA under the Obama administration in order to boost the productivity of both industries.

6.8.6 The Obama Administration

The agreements pertaining to climate change were signed to be ratified by the United States of America under the Obama administration. During his tenure as President, Obama implemented the Clean Power Plan to combat climate change. Through this plan, Obama ensured that during his tenure, carbon emissions decreased by 9 per cent, while the economy grew by 10 per cent.[78] Obama's Clean Power Plan ensured that carbon emissions from power plants throughout the country while maintaining their affordability and energy reliability.[79] Under the auspices of the US Environmental Protection Agency, Obama wanted to fully implement the Clean Power Plan by 2030 to protect Americans from harmful emissions and pollution.[80] While Obama envisaged this national plan for 2030, he urged states to implement their own reduction plans through the course of the years to ensure the required cuts by 2030.[81] By reducing emissions, it is also essential that the power plants do not hamper their reliability to provide electricity nationally.[82] Some of the major issues plaguing the US today are heat waves, droughts and the increased intensity of hurricanes.[83] In addition to ensuring lesser emissions from power plants, the Obama administration pushed for technological advancements in the automobile industry to reduce emissions. The administration actively pushed manufacturers to continue to push boundaries and also continue with the current trend of manufacturing cars that utilise electricity and zero emission vehicles.[84] In order to facilitate a smooth transition in 2030, the US Environmental Protection

Agency gave each state a goal a goal to reduce their emissions. The individual states were then allowed to decide how to meet these goals. In order to ensure the safety of Americans, the Obama administration went to great lengths to ensure emissions were gradually reduced by empowering the EPA to set certain goals. President Obama himself knew of the dangers of climate change and knew the global implications and ramifications of not ratifying certain climate change agreements. He also encouraged manufacturers to use renewable fuel sources, instead of fossil fuels. Moreover, he was instrumental in bring the US, India and China, the three biggest global emitters, to ratify the Paris Agreement. Furthermore, he also understood the security implications of unchecked emissions. While most of us focus more on global security concerns pertaining to climate change, he used the opportunity to concentrate on honing US policies to tackle climate change first in the US and then globally. During his tenure as President, it can be argued that he was looked at as a beacon of hope to tackle climate change. NGOs and other supporters of his policies to tackle climate change were vocally supportive of certain measures undertaken by his government. With the election of President Trump, many Green Party supporters and NGOs are apprehensive that he would overturn the policies implemented by the Obama administration to help tackle climate change.

6.8.7 The Trump Administration

With the end of the Obama Administration, President Trump is the current Commander in Chief of the United States of America. Throughout his presidential campaign, President Trump has routinely denied climate change and continued to do so since taking office. Recently, President Trump nominated Scott Pruitt to lead the Environmental Protection Agency. Pruitt's nomination was marred by his controversial stand on climate change. He is widely known to be a staunch legal opponent of the EPA.[85] As a sceptic of climate change, Scott Pruitt has questioned the very rationale of human caused climate change, in an attempt to deregulate the EPA rules.[86] Having served as the governor of Oklahoma, Pruitt spearheaded a number of lawsuits against the EPA (with little to no success), to try and halt the EPA rules on mercury pollution from coal plants.[87] After his nomination as the head of the EPA, nearly 800 former EPA employees signed a letter opposing his confirmation, owing to his alleged close ties with the coal and gas industry in Oklahoma.[88]

President Trump wanted a nominee who would scale down President Obama's environmental policies. Being tasked with enforcing laws ranging from air pollution to chemical safety, the EPA is also guided by major statutes like the Clean Air Act and Clean Water Act,[89] which need to be regularly updated to keep up with new and evolving science on public health and pollution. It is therefore the EPA's job to ensure the safety of air and water, amongst other things. Today, the EPA is the key US Agency dealing with climate change. Furthermore, in 2007, the Supreme Court ruled that the EPA has the authority to regulate greenhouse gases under the Clean Air Act, if they deemed a threat to public safety.[90] With Pruitt's confirmation, he would start by weakening and rewriting some of the policies, implemented by the EPA under Obama.[91] In addition to this, he will also delay implementing new policies which would help regulate greenhouse emissions from automobiles, manufacturing plants, refineries, chemical plants and others.[92] In addition to this, Pruitt might also be instrumental in reducing the EPA's budget.[93]

Through the Obama administration, the world took steps to try and counter climate change signed and ratified a number of agreements to this accord, with Obama's goal of cutting greenhouse emissions by at least 26 per cent by 2020[94] setting certain rules of the EPA in motion. However, under the Trump administration and under Pruitt's leadership, the EPA faces the daunting task of repealing some, if not most, of its policies with regards to the Clean Air Act and the Clean Water Act.[95] One of the main concerns of Pruitt's confirmation as head of the EPA is if he would undermine the Paris Agreement, where each country has a goal to achieve to reduce its emissions.[96] This being said, it is hard to predict the decisions that will be implemented by Pruitt and the Trump administration. While the Paris Agreement was heralded as a global success, many believed that President Trump will try and find a way to pull out of the treaty. On June 1, 2017, United States President Donald Trump announced that the US would cease all participation in the 2015 Paris Agreement on climate change mitigation. Trump stated that "The Paris accord will undermine (the US) economy," and "puts (the US) at a permanent disadvantage." During the presidential campaign, Trump had pledged to withdraw from the pact, saying a withdrawal would help American businesses and workers. Trump stated that the withdrawal would be in accordance with his America First policy. Throughout his campaign, Trump had been vocal about his criticisms of the UN. In the future, in order to block further deals and agreements on climate change, Trump could even pull out of the UNFCCC.[97]

6.9 Comparative Analysis of Indo-US Response to Climate Change

In September, 2014, the US and India pledged to strengthen and expand the highly successful US-India Partnership to Advance Clean Energy (PACE). Since then, the two countries have made important new activities to advance both countries ambitious clean energy objectives and Nationally Determined contributions. The United States had appeared to be careful not to be confrontational with India ahead of the Paris climate conference, after New Delhi reacted sharply to Secretary of State John Kerry's statement that India would be a "challenge" in negotiations.[98] A State Department spokesperson refused to elaborate what the secretary of state meant by "challenges" and instead chose to partnering in combating climate change. In November and December 2015, the 21st conference of the parties to the United Nations Framework Conventions on Climate Change (UNFCCC COP21) took place in Paris. UNFCCC is an international environmental agreement on climate change of which there are 195 states parties, including the UK.

The key points of the Paris agreement are summarised below. The agreement is due to come into force in 2020.

6.9.1 Mitigation: Reducing emissions Governments agreed

- to a long-term goal of keeping the increase in global average temperature to well below 2°C above pre-industrial levels;
- to aim to limit the increase to 1.5°C, since this would significantly reduce risks and the impacts of climate change;
- on the need for global emissions to peak as soon as possible, recognising that this will take longer for developing countries;

- to undertake rapid reduction thereafter in accordance with the best available science.

6.9.2 Transparency and global stocktake Government agreed to

- come together every 5 years to set more ambitious targets as required by science;
- report to each other and the public on how well they are doing to implement their targets;
- track progress towards the long-term goal through a robust transparency and accountability system.

6.9.3 Adaptation

Governments agreed to:

- Strengthen society's ability to deal with the impacts of climate change;
- Provide continued and enhanced international support for adaptation to developing countries.

6.9.4 Loss and Damage

The agreement also:

- recognizes the importance of averting, minimising and addressing loss and damage associated with the adverse effects of climate change;
- acknowledges the need to cooperate and enhance the understanding, action and support in different areas such as early warning systems, emergency preparedness and risk insurance.

6.9.5 Support

- The EU and other developed countries will continue to support climate action to reduce emissions and build resilience to climate change impacts in developing countries.
- Other countries are encouraged to provide or continue to provide such support voluntarily.
- Developed countries intend to continue their existing collective goal to mobilise USD 100 billion per year until 2025 when a new collective goal will be set.

6.10 US and India's Environment Policies after the Paris Agreement

From the beginning India has played an active role in Climate change debate, which over the period of time India has asserted its interest in climate change politics. After Paris agreement, India launched the International Solar Alliance, and India is diversifying its energy mix with, more focus on renewable energy source. After back out by USA, India still pledged to uphold the climate change commitment, although, it would be difficult to achieve all NDC. The Climate change debate will take an unexpected turn once China gets centre stage in climate change debate. After president Obama left office, there has been big debate on US environment policy in President Trump's regime. One of the Trump's administrative priorities was to roll back from the Obama-era government environmental regulations. With the denial of Paris agreement, Trump said that "I was elected to represent the citizens of Pittsburgh, not Paris," At least the mayor of Pittsburgh tried to reply via Twitter that ""I, Bill Peduto, assure you that we will follow the guidelines of the Paris agreement for our people, our economy and future." This is not the only case of conflict after Trump took office as the US president. There are plenty of different cases of laws and policy changes in US Environment Policy.

The Trump's policy list is really long in shaping new environment policy of his regime. The Paris climate deal was also one of these environment policy reforms in US. Trump is seeing it as unfairly pushing the US while other countries like China and India or less developed countries continue with industrialization through oil and coal. He found it unfair to the American coal industry and also with his idea of America First. He even declared that America will not contribute to the UN Green Climate Fund. This fund was developed for the projects which help developing countries that don't rely on fossil fuels. Under the presidency of Obama, the government came with the Clean Power Plan. This plan was followed by the Paris Climate Deal to keep US in the line of completion of it. It proposed the guideline for the state by state emissions reduction. But the plan was stuck at the courts from the beginning in 2015. The new administration and head of EPA is looking to rescind it. The Budget of the EPA was cut down around more than 30 percent.

Most of the important members of the EPA are taking buyouts. The US government gave green signals to the new exploratory drilling operations through the GOP tax bill. There were not any drilling operations in the Alaskan National Wildlife Refuge since 1980s. Oil is flowing through the pipelines in the US again in Trump's administration. The Dakota Access Pipeline is active since the passed by the Trump administration.

The group of 300 US scientists and 13 federal agencies came out with an assessment of climate change on 23 November, 2018. They found that America was already experiencing wildfires, storms, heatwaves and with the climate change it would further disrupt many areas of life in the future. The US National Climate change assessment of climate change and rising temperatures, warns in the report, about hundreds of billions of dollars lost, heavily effect of the agriculture with crop failures, expanding wildfires, different health issues, and altered coastlines. The report claimed that rise in temperature could estimate up to tenth of US GDP with the rising sea levels and storms. Because of heatwaves, deferent diseases like asthma and pulmonary disease will increase and will affect thousands of people in the US. Harvard University professors, Devid Cutler and Francesca Dominici, analysed the EPA policies on toxic chemicals and air pollution. They concluded that because of air pollution and toxic chemicals the death figure will increase around 80,000. The EPA answered it by saying that it is not a scientific report but it is a more politically motivated one. The EPA head was changed to implement of Trump's environment policies. The new head, Scott Pruitt, denied the existence of global warming in the past. He worked against Obama's initiative of Clean Water Rule. Lukas Ross, an environment activist, said in his interview that Trump is trying to appoint his cabinet and government agencies with undeserved corporate cronies whose bad impact can easily affect EPA policies. The appointments of Scott Pruitt for the head of EPA and Ryan Zinke at the department of the Interior are the best examples for the same. The change in the environment polices will impact generations of the US citizen. It will not be easy for the future generation to fix the harm which has been done by the Trump's administration. Since Scott Pruitt took charge of the head of EPA, many programs and policies were shut down. Many prominent personalities and environment experts resigned from the EPA. It won't be easy in the future to establish this institution with such experts.

6.11 Indo-US Comparative Analysis

Compared to other countries, India is in a unique situation. India today is on a trajectory towards industrial development. In addition to this, India is stuck between nations that industrialised a long time ago and can now afford renewable sources of energy, and countries which are still reliant on pre-industrial low carbon practices to gain their livelihood.[99] Despite being known as an economic weight and an emerging power, India still bears the hallmarks of a least developed country in its villages and communities.[100] Furthermore, the sheer size of India's population makes India's choices about development and climate and energy carry global consequences for its decisions. One of India's main problem is the fact that several generations of Indians are on the cusp of prosperity and development, provided growth is powered by cheaper energy.[101] India today bears a striking resemblance to the US, and other industrialised countries in the 19th and 20th centuries, as well as China in the recent past.[102] At a time when India is getting industrialised, the world is willing to take multilateral action against climate change. For the US, having India on board for any climate change agreement is beneficial. India would neither be a signatory, nor would ratify an agreement that would hamper its growth, thereby giving more leverage to the US to continue its greenhouse emissions.

India is slowly making advances in the renewable energy sector. Today, India is the world's largest biomass, the third largest solar, and the fourth largest wind energy producer based on 2015 ranking.[103] The combined generation from solar, wind, small hydro, biomass, bagasse, and other sources stood at 10.647 billion units (BU) in December 2019. Among renewable sources, wind-generated the highest amount of power followed by solar at 3,943 million units (MU) and 3,932 MU, respectively, in December 2019.[104] In January 2020, India's power generation from the sources of renewable energy reached at 10.325 billion units (BU) which was higher in comparison to the previous year that stood 9.433 BU as per the Central Electricity Authority (CEA) report.

The US, on the other hand, has pledged to reduce its carbon footprint by slowly reducing emissions from manufacturing plants by switching from fossil fuels to renewable sources of energy. India's main point of contention for reducing and eventually stopping using coal for its power dependence, is creating a climate framework that would create meaningful funding and technology transfer to help accelerate India's move from coal consumption to renewable sources.[105] If India wants to eradicate poverty, develop industrially and stop mass urban migration, it will need to deliver cheap energy to its citizens, in its villages and other communities. Despite India's want on need to be a developed country, its emissions are far lesser than the US. The US carbon emissions for the year 2011 accounted for 16 per cent of total global emissions, while India's accounted for 6 per cent.[106] Therefore, the policies between the two countries will forever be changing, as each country faces different threats due to climate change. The Paris Agreement was one of the only agreements where both the countries agreed to tackle climate change together. In accord with the Paris Agreement, the US has implemented the Clean Air Act and the Clean Water Act through the EPA to reduce emissions. India on the other hand, implemented eight national missions,[107] national solar mission, national mission for enhanced energy efficiency, national mission on sustainable habitat, national water mission, national mission for sustaining the Himalayan ecosystem, national mission for a green India, national mission for sustainable agriculture, national mission on strategic knowledge on climate change.[108]

In order to be implemented, the missions are first institutionalised under their respective ministries, and then the respective ministries implement their respective missions, under specific plans like the 11th and 12th Plans for 2015-16 and 20-17 respectively.[109] While the Indian government institutionalizes its missions, the US government gives the EPA certain autonomy to enact acts as it sees fit under the auspices of its head and the President. In addition to this, the EPA also has the power to nationally assess greenhouse emissions and control them accordingly. The Indian missions on the other hand, are bound by the rules and regulations of their respective ministries are not allowed to operate beyond their set purview. The striking difference between the US and India, is the fact that courts in the US can intervene on behalf of green groups and the EPA, while in India, the process of going through the courts could prove to be a difficult course of action. Under the Obama administration, the US and India had similar ideologies and processes to combat climate change, but under President Trump, it is difficult to assess what policies will be implemented. While both the countries face almost similar threats with regard to climate change, they both have their own methods to try and tackle climate change. The stark difference between the Indian government and the US administration is that the Indian government is still focused on developing its industrial economy on fossil fuels before switching to renewable energy, while the US is trying to reduce its emissions by switching from fossil fuels to renewable sources. With regard to human security, India faces a crisis of mass urban migration with its industrialisation process. The number of people moving from villages to the cities is increasing with each passing year. The US, faces issues from chemical pollution, air pollution, smog and other emissions. Therefore, while both the countries do face their own problems, each country tackles climate change through their own means.

With regards to climate change, there are stark difference between India and the US. The US tries to tackle climate change with the EPA, while India does the same with the National Action Plan for Climate Change (NAPCC).[110] The two countries are together responsible for an increased number of greenhouse gases emissions. Therefore, through their respective agencies, both countries have implemented policies to help combat climate change. However, it has been tough for both countries to understand the brevity of climate change. While the US has not ratified any treaty or agreement till the Paris Agreement, India has ratified both, the Kyoto Agreement, as well as the Paris Agreement.

With the election of Indian Prime Minister, India is slowly making advances in the renewable energy sector. Today, India is the world's largest biomass, the third largest solar, and the fourth largest wind energy producer[111] The US, on the other hand, has pledged to reduce its carbon footprint by slowly reducing emissions from manufacturing plants by switching from fossil fuels to renewable sources of energy. India's main point of contention for reducing and eventually stopping using coal for its power dependence, is creating a climate framework that would create meaningful funding and technology transfer to help accelerate India's move from coal consumption to renewable sources.[112] If India wants to eradicate poverty, develop industrially and stop mass urban migration, it will need to deliver cheap energy to its citizens, in its villages and other communities. Despite India's want on need to be a developed country, its emissions are far lesser than the US. The US carbon emissions for the year 2011 accounted for 16 per cent of total global emissions, while India's accounted for 6 per cent.[113] Therefore, the policies between the two countries will forever be changing, as each country faces different threats due to climate change. The Paris Agreement was one of the only agreements where both the countries agreed to tackle climate change together. In accord with the Paris Agreement, the US has implemented the Clean Air Act and the Clean Water Act through the EPA to reduce emissions. India on the other hand, implemented eight national missions,[114] national solar mission, national mission for enhanced energy efficiency, national mission on sustainable habitat, national water mission, national mission for sustaining the Himalayan ecosystem, national mission for a green India, national mission for sustainable agriculture, national mission on strategic knowledge on climate change.[115] In order to be implemented, the missions are first institutionalised under their respective ministries, and then the respective ministries implement their respective missions, under specific plans like the 11th and 12th Plans for 2015-2016 and 2016-2017 respectively.[116]

While the Indian government institutionalizes its missions, the US government gives the EPA certain autonomy to enact acts as it sees fit under the auspices of its head and the President. In addition to this, the EPA also has the power to nationally assess greenhouse emissions and control them accordingly. The Indian missions on the other hand, are bound by the rules and regulations of their respective ministries are not allowed to operate beyond their set purview. The striking difference between the US and India, is the fact that courts in the US can intervene on behalf of green groups and the EPA, while in India, the process of going through the courts could prove to be a difficult course of action. Under the Obama administration, the US and India had similar ideologies and processes to combat climate change, but under President Trump, it is difficult to assess what policies will be implemented.

The US-India strategic partnership has become especially significant in the context of the United States rebalancing policy in the Asia-Pacific region. Recognizing the profound threat of global climate change, the two sides stressed the importance of working together and with other countries to conclude an ambitious climate agreement. It was realised that the goal would require concerted action by all countries and the international community.[117] A strengthened and expanded US-India partnership is significant to advance clean energy, Accelerating Renewable Energy Deployment, Grid integration, Solar Resource Data, US-India clean Energy Finance Task Force, New Clean Energy Finance initiative and US-India Catalytic Solar Finance Programme. For energy efficiency-with the expansion of historical collaboration, the United States and India are renewing the Energy Conservation Building code and Supporting Comprehensive State-level policies for the current framework, while making significant progress on space cooling collaboration.[118] While the relationship between the two countries thrived under the Obama administration, India still does not have a clear policy plan regarding climate change from the Trump administration. India continues to work on its missions to tackle climate change and anticipates to work closely with the US to tackle climate change.

NOTES

1. India-United States Relations, Accessed through https://www.mea.gov.in/Portal/ForeignRelation/USA_15_01_2016.pdf on 2 December 2016.
2. India-United States Relations, Accessed through https://www.mea.gov.in/Portal/ForeignRelation/USA_15_01_2016.pdf on 25 November2016.
3. India-United States Relations, Accessed through https://www.mea.gov.in/Portal/ForeignRelation/USA_15_01_2016.pdf on 25 November 2016.
4. Pande, Varad, Changing the Climate of the India-U.S. Strategic Partnership 9 October 2013, Accessed through http://www.business-standard.com/article/opinion/varad-pande-changing-the-climate-of-the-india-us-strategic-partnership-113101200638_1.html on 2 December2016.
5. Joint Statement: The United States and India: Enduring Global Partners in the 21st Century: Accessed through https://obamawhitehouse.archives.gov/the-press-office/2016/06/07/joint-statement-united-states-and-india-enduring-global-partners-21st on 8 July 2019.
6. Ibid.
7. US State Department, Accessed throughwww.state.gov/r/pa/prs/ps/2015/09/247192.html on 3 November2016.
8. US State Department, Accessed through www.state.gov/r/pa/prs/ps/2015/09/247192.html on 5 November2016.
9. White House Press Office, Accessed through www.whitehouse.gov/the-press-office/2016/06/07on 13 November 2016.
10. Ibid.
11. U.S.-India Relations 1947-2020 Accessed Through, https://www.cfr.org/timeline/

us-india-relations on 10 November 2020.

12. Burns R. Nicholas, America's Strategic Opportunity with India: The New U.S.-India, Partnership Council on Foreign Relations, Foreign Affairs, Vol. 86, No. 6 Nov.-Dec., 2007, pp.131-46, under 'Missed Opportunities'.
13. Ibid.
14. Feigenbaum, Evan A., India's Rise, America's Interest: The Fate of the U.S.-Indian Partnership, Council on Foreign Relations, Foreign Affairs, Vol. 89, No. 2 March/ April 2010,pp. 76-91.
15. Ibid.
16. Ibid.
17. Devenport, Kelsey, Nuclear Security Summit at a Glance Accessed through https:/ /www.armscontrol.org/factsheets/NuclearSecuritySummit#:~:text= The%20first%20Nuc lear%20Security%20Summit,and%20the%20European% 20Union%20attendedon 20 May 2019.
18. Strategic Energy Partnership emerged as cornerstone of expanding India-US ties: Envoy, accessed through https://www.financialexpress.com/defence/strategic-energy-partnership-emerged-as-cornerstone-of-expanding-india-us-ties-envoy/ 2093162/ on 19 October 2020.
19. India-United States Relations, Accessed through https://www.mea.gov.in/Portal/ ForeignRelation/USA_15_01_2016.pdf on 2 December 2016.
20. India-United States Relations, Accessed through https://www.mea.gov.in/Portal/ ForeignRelation/USA_15_01_2016.pdf on 2 December2016
21. Chatterjee, Anup, Global Warming & Climate Change, Global Publications, p79, Delhi 2010.
22. Ibid.
23. Ibid,p.80.
24. Ibid.
25. Ibid.
26. Hardy T. John, Climate Change Causes, Effects, and Solutions, John Wiley & Sons Ltd, The Atrium, Southern Gate, Chichester 2003, p.221.
27. Chatterjee, Anup, "Global Warming & Climate Change", Global Publications, p.81, Delhi 2010.
28. Thaker J and Leiserowitz A (2014), Shifting discourses of climate change in India, Climatic Change. Accessed through: http://environment.yale.edu/climate-communication-Thaker_and_Leiserowitz_(2014)_Shifting_discourses_of_climate_change_in_India.pdf on 22 November 2016. 135IEA International Energy Agency (2013) CO_2 emissions from fuel combustion– highlights. Accessed through: http:/ /www.iea.org/co2highlights/co2highlights.pdf on 22 November 2016
29. Agarwal, A.; Narain, S., Global Warming in an Unequal World. A case of environmental colonialism. Accessed through http://cseindia.org/ challenge_balance/readings/GlobalWarming%20Book.pdf on 22 November2016
30. Singh Jasjit, "India's Strategic and Security Perspective", Vol.XIII, Issue No. 5, August 1990, pp. 475-500
31. Agarwal, A.; Narain, S., Global Warming in an Unequal World. A case of environmental colonialism. Accessed through http://cseindia.org/ challenge_balance/readings/GlobalWarming%20Book.pdf on 22 November 2016.

32. Dubash N.K. Copenhagen: Climate of mistrust, Economic & Political Weekly 64, 8–11. Accessed through http://re.indiaenvironmentportal.org.in/files/COP.pdf on 24 November 2016.
33. Brodie J (2009) *From Social Security to Public Safety: Security Discourses and Canadian Citizenship*. University of Toronto Quarterly, Volume 78, Number 2, pp. 687-708, 2009.
34. Ibid.
35. Koithara Verghese, "Society, state, and security the Indian Experience", Sage Publication New Delhi, 1999,p. 20.
36. Ibid.,p 22.
37. Detraz N and Betsill M., Climate Change and Environmental Security: For Whom the Discourse Shifts. International Studies Perspectives 2009 10, 303–320. Accessed through: http://graduateinstitute.ch/files/live/sites/iheid/files/sites/political_science/shared/political_science/ 9957/Detraz&Betsill%202009.pdf on 25 November 2016.
38. Ibid.
39. Thaker J and Leiserowitz A (2014) Shifting discourses of climate change in India, Climatic Change. Accessed through: http://environment.yale.edu/climate-communication-OFF/files/Thaker_and_Leiserowitz_(2014)_Shifting_discourses_of_climate_change_in_India.pdf on 22 November 2016.
40. Ereaut, G.,Segnit, N., Warm words: How are we telling the climate story and can we tell it better? Technical Report, Institute for Public Policy Research, London, 32. 2006
41. Risbey J., The new climate discourse: Alarmist or alarming? *Global Environmental Change*, Vol. 18, pp. 26-37, 2008.
42. Rajamani, Lavanya, India and Climate Change: What India Wants, Needs, and Needs to Do, India Review, vol.8, no. 3, July-September 2009, pp.340-74.
43. Ibid.
44. Rajamani, Lavanya, "India and Climate Change: What India Wants, Needs, and Needs to Do", India Review, vol.8, no. 3, July-September, 2009, pp.340-74.
45. By the Numbers: The US's warmest year yet. Accessed through http://www.popsci.com/science/article/2013-01/numbers-uss-warmest-year-yet-infographic on 20 January2017.
46. NOAA Climate Services. Accessed through https://www.ncdc.noaa.gov/oa/about/response-v2.pdf on 20 January 2017.
47. The Copenhagen Accord: Foundations for International Action on Climate Change. Accessed through https://www.nrdc.org/sites/default/files/Copenhagen%20accord.pdf on 20 January 2017.
48. What Cancun means for China and the US. Accessed through http://www.wri.org/blog/2010/12/what-cancun-means-china-and-us on 20 January 2017.
49. Ibid.
50. Durban: At last, US, India and China agree to act on climate. Accessed through http://www.theweek.co.uk/environment/climate-change/43518/durban-last-us-india-and-china-agree-act-climate on 20 January2017.
51. Ibid.
52. The UNFCCCs Durban Platform. Accessed through http://www.hfw.com/

downloads/Client%20Brief%20-%20UNFCCCs%20Durban%20%5BA4%204pp %5D% 20January%202012.pdf on 25 January 2017.

53. UN Climate talks end with late deal. Accessed through http://www.bbc.com/news/science-environment-16124670 on 25 January 2017.
54. Ibid.
55. The Paris Agreement explained: Accessed through http://www.abc.net.au/news/2016-12-09/the-paris-agreement-explained/8107100on 17 January2017.
56. Ibid.
57. Ibid.
58. Ibid.
59. United States Environmental Protection Agency Our Mission and what we do: Accessed through https://www.epa.gov/aboutepa/our-mission-and-what-we-do on 17 January 2017.
60. Ibid.
61. Ibid.
62. Summary of the Clean Air Act: Accessed through https://www.epa.gov/laws-regulations/summary-clean-air-act on 17 January 2017.
63. Ibid.
64. Air Enforcement: Stationary Sources: Accessed through https://www.epa.gov/enforcement/air-enforcement on 17 January 2017.
65. Ibid.
66. Ibid: Mobile Sources
67. Ibid.
68. Ibid.
69. Ibid.
70. Ibid.
71. Summary of the Clean water Act: Accessed through https://www.epa.gov/laws-regulations/summary-clean-water-act on 17 January2017.
72. Clean Water Act Compliance Monitoring: Accessed through https://www.epa.gov/compliance/clean-water-act-cwa-compliance-monitoring on 17 January 2017.
73. Ibid.
74. Ibid.
75. Ibid.
76. Ibid.
77. Ibid.
78. A historic commitment to protecting the environment and addressing the impacts of climate change. Accessed through https://obamawhitehouse.archives.gov/the-record/climate on 1 February 2017.
79. Fact Sheet: Overview of Clean Power. Accessed through https://www.epa.gov/cleanpowerplan/fact-sheet-overview-clean-power-plan on 1 February 2017.
80. Ibid.
81. Ibid.
82. Ibid.
83. Learn about carbon pollution from power plants. Accessed through https://www.epa.gov/cleanpowerplan/learn-about-carbon-pollution-power-plants#what

on 1 February2017.
84. A historic commitment for protecting the environment and addressing the impacts of climate change: The work ahead. Accessed through https://obamawhitehouse.archives.gov/the-record/climate on 1February2017.
85. Senate confirms Scott Pruitt as EPA head: Accessed through https://www.nytimes.com/2017/02/17/us/politics/scott-pruitt-environmental-protection-agency.html?_r=0 on 17 February2017.
86. Ibid.
87. What Scott Pruitt's confirmation means for the EPA: Accessed through http://www.vox.com/science-and-health/2017/2/17/14630614/scott-pruitt-epa-climate on 17 February 2017.
88. Ibid.
89. Ibid.
90. Ibid.
91. Ibid.
92. Ibid.
93. Ibid.
94. Ibid.
95. Ibid.
96. Ibid.
97. Ibid.
98. Varghese K. George, India a partner in Climate Change fight: US, Accessed through www.thehindu.com/news/international/india-partner-in-climate-change-talks-says-us/article7930177ece on 29 November2015.
99. An 'Indian exception' and India-US partnership on climate change: Accessed through https://www.brookings.edu/blog/planetpolicy/2015/01/12/an-india-exception-and-india-u-s-partnership-on-climate-change/ on 12February2017.
100. Ibid.
101. Ibid.
102. Ibid.
103. Ibid.
104. India's renewable energy generation grows 9.46 per cent in Jan 2020. Accessed through https://energy.economictimes.indiatimes.com/news/renewable/indias-renewable-energy-generation-grows-9-46-per-cent-in-jan-2020/74442008 on 2 March 2020 on 10 May 2020.
105. Ibid.
106. Global Greenhouse Gas emissions data. Accessed through https://www.epa.gov/ghgemissions/global-greenhouse-gas-emissions-data on 13February2017.
107. National Action Plan on Climate Change: Accessed through http://www.moef.nic.in/modules/about-the-ministry/CCD/NAP_E.pdfp. 2, on 13 February2017.
108. Ibid, pp. 3-5.
109. Ibid, pp. 5-6.
110. Comparative Analysis of developing country national climate change strategies: Accessed through http://www.usclimatenetwork.org/resource-database/WRI%20Comparative%20Analysis%20OF%20Developing%20Country%20National

%20Climate%20 Change%20Strategies.pdf on 12February, 2017.

111. Ibid.
112. Ibid.
113. Global Greenhouse Gas emissions data: Accessed through https://www.epa.gov/ghgemissions/global-greenhouse-gas-emissions-data on 13 February 2017.
114. National Action Plan on Climate Change: Accessed through http://www.moef.nic.in/modules/about-the-ministry/CCD/NAP_E.pdf p. 2, on 13 February2017.
115. Ibid, pp. 3-5.
116. Ibid, pp. 5-6.
117. US State Department, Accessed throughwww.state.gov/r/pa/prs/ps/2015/09/247192.html on 3 November2016.
118. White house press office, Accessed through www.whitehouse.gov/the-press-office/2016/06/07 on 13 November2016.

7

Politics of Global Actors and Climate Change

7.1 International Response to Climate Change and Role of International Agencies

The climate system should be protected for the benefit of present and future generations of humankind, on the basis of equity and in accordance with their common but differentiated responsibilities and respective capabilities. Climate change is a global problem and its impacts and consequences are often felt far away from the origin of GHG emissions. For effective mitigation and adaptation international co-operation will prove to be a vital tool. Climate change has resulted in global warming which is the most serious environmental problem that the world is facing. The threats in terms of the expected impacts on ecosystems, livelihoods, human health and food security are enormous. While the biggest contributors to global warming are the industrialized countries due to the increasing greenhouse gas emissions, the greatest impacts of climate change will be felt first, and seriously, by the poorest of under developing countries of the world. The challenge of climate change resulting in global warming and aggravating the problems with all its consequences requires prompt and early international response to address climate change. The IPCC's AR4 put forth a set of criteria for assessing means of international cooperation which include the following:

- Environmental effectiveness
- Cost-effectiveness
- Distributional considerations
- Institutional feasibility

The timeline for the various international responses to climate change is summarized in Table 3.

Table 3: Timeline Detailing the International Responses to Climate Change

Year	*International Response*
1979	• First World Climate Conference took place.
1988	• The Intergovernmental Panel on Climate Change (IPCC) was setup.
1990	• IPCC's first assessment report was released. • IPCC and world climate conference called for a global treaty on climate change. • United Nations General Assembly negotiations on a framework convention began.
1991	• First meeting of the Intergovernmental Negotiating Committee (INC) took place.
1992	• The INC adopted UNFCCC (United Nations Framework Convention on Climate Change) text. • At the Earth Summit in Rio, UNFCCC was opened for signature along with its sister Rio Conventions, UNCBD and UNCCD.
1994	• UNFCCC entered into force
1995	• First Conference of the parties (COP 1) took place in Berlin.
1996	• The UNFCCC Secretariat was set up to support action under the convention.
1997	• Kyoto Protocol was formally adopted in December at COP 3.
2001	• IPCC's Third Assessment Report was released. • Bonn Agreements were adopted based on the Buenos Aires Plan of Action of 1988. • Marrakesh Accords were adopted at COP 7, detailing rules for implementation of Kyoto Protocol, setting up new funding and planning instruments for adaptation and establishing a technology transfer framework.
2005	• Entry into force of Kyoto Protocol. • The first meeting of the parties to the Kyoto Protocol (MOP 1) took place in Montreal. • In accordance with Kyoto Protocol requirements, Parties launched negotiations on the next phase of the protocol under the Ad Hoc Working Group on Further Commitments for Annex 1 Parties under Kyoto Protocol (AWG-KP). • What was to become the Nairobi Work Programme on Adaptation (it would receive its name in 2006, one year later was accepted and agreed?
2007	• IPCC's Fourth Assessment Report was released. • Climate Science entered into popular consciousness.

Year	International Response
	• At COP 13, Parties agreed on Bali Road Map, which charted the way towards a post-2012 outcome in two work streams: the AWG-KP and another under the Convention, known as the Ad-Hoc Working Group on Long-Term Cooperation Action under the Convention.
2009	• Copenhagen Accord drafted at COP 15 in Copenhagen. • This was taken note of by the COP. • The Countries later submitted emissions reductions pledges or mitigation action pledges, all non-binding.
2010	• Cancun Agreements were drafted and largely accepted by the COP, at COP 16
2011	• The Durban Platform for Enhanced Action was drafted and accepted by the COP, at COP 17.
2012	• The Doha Amendment to the Kyoto Protocol was adopted by the CMP at CMP 18.

Source: IPCC.

According to IPCC WGIII AR5 the principles of climate change policy relevant for international cooperation can be grouped into several broad categories as follows:

1. The principle of maximizing global net benefits makes the trade-off between aggregate compliance costs and aggregate performance benefits explicit. The principle also incorporates the notion of maximizing co-benefits of climate action.[1,2,3,4,5]
2. Equity is a principle that emphasizes distributive justice across and within countries and across and within generations.[6,7,8,9,10] It includes evaluating the procedures used to reach an agreement as well as the achieved outcomes. This principle may also apply in a broader assessment of well-being.[11,12]
3. Third, the principle of precaution emphasizes anticipation and prevention of future risks, even in the absence of full scientific certainty about the impacts of climate change.[13,14,15] Some see precaution as a strategy for effective action across diverse uncertain scenarios, although the application of precaution varies across risks and countries.[16]
4. Fourth, the principle of sustainable development, broadly defined, emphasizes consideration of the socioeconomic needs of future generations in making decisions about current resource use.[17,18]

Keohane and Young showed that international institutions like the treaties, organizations, and behavioural norms—aid cooperation by reducing those transaction costs, focusing efforts on particular solutions, creating reputational risks for failure and the like[19,20] According to Victor D.G., there are four aspects of climate damages that can affect whether and how nations are likely to mobilize to address the dangers of changing climate:

Aspect 1: Increasing capacity of societies to "climate proof" themselves. A very small fraction of economic activity depends directly on the weather and climate; human capacity to adapt to changing weather, such as by building dikes and irrigation systems, has risen sharply in the last century and shows no sign of exhaustion.

Aspect 2: Climate change takes place over a period of time. The calculation of reluctant developing countries is, perhaps, entirely rational. Combating global warming would require them to incur possibly a substantial cost in controlling emissions today for quite uncertain benefits in the future. By contrast, they could invest the same resources today in development, which automatically will improve their future capacity to adapt if climate should change (while also achieving many other benefits from development).

Aspect 3: The third aspect of climate science is that in fact some societies might welcome a dose of new climate up to a point. The most famous of these is Russia (and more generally the former Soviet Union), where agriculture and forestry which are the most climate-sensitive of economic activities probably stand to gain from the longer growing seasons that accompany warmer weather.

Aspect 4: It includes scenario for changing climate that all countries have an unequivocal interest in avoiding: abrupt (or so-called "catastrophic") climate change, such as would be caused if any of the following were to occur: the sea level were to rise a meter or more over just a few decades; the world's climate were to "flip" to a different regime; warming destabilized large amounts of methane currently locked in the permafrost (which would trigger still more warming, as methane is a strong greenhouse gas); or the ecological integrity of the Amazon or some other huge swath of the world's ecosystems were undermined.[21]

7.2 How are Global Actors Dealing with Climate Change and Security Change?

Efforts to mainstream climate change adaptation into national development planning are still at a relatively early stage in many countries. Countries are increasingly requesting the United Nations Development Programme—United Nations Environment Programme Poverty-Environment Initiative (PEI) to tackle the mainstreaming of climate change adaptation into their national development planning, as part of broader poverty-environment mainstreaming efforts. It is essential that adaptation policies or strategies are formulated as part of broader policies for development. Implementing specific adaptation measures (geared to specific problems, sectors or population groups) may be effective in certain circumstances, but in the long run, a project-based approach to adaptation planning and financing may not produce the scale of results that is needed. In this respect, efforts to formulate national adaptation policies or climate change strategies will need to be supported by a cross-cutting, integrated policy approach.

7.3 United Nation Organization (UNO) and Climate Change

United Nations Secretary-General Ban Ki-moon made progress on global climate change one of his top priorities, and he has been working with countries to help them reach an agreement since he took office in 2007. Both the global negotiations and the scientific assessment of the challenge are led by the UN. International efforts to deal with climate change began at the Rio Earth Summit in 1992, with the adoption of the UN Framework Convention on Climate Change a treaty signed and ratified by the United States. This agreement makes all countries responsible for working to avoid "dangerous human interference" with the climate system. It was based from its very first words on the idea that "change in the Earth's climate and its adverse effects are a common concern of humankind."

In 1997, an implementation agreement of the Framework Convention was adopted. The Kyoto Protocol, which covered 170 countries, committed developed countries to reducing their greenhouse gas emissions by 5 to 7 percent from 1990 levels by 2012. The US signed but did not ratify the Kyoto Protocol. International negotiators have pledged to reach a new global agreement on climate change at a conference in December 2015 in Paris. They are taking a new approach based on voluntary national commitments to action a "race to the top" in climate-friendly energy solutions. The shift to a low-carbon economy is occurring everywhere. Countries, states, cities, and companies are taking action, partly out of concern about climate change, but also because such action is in their own economic interest either directly (through money-saving efficiency measures like LED street lights) or indirectly by saving lives now cut short by air pollution). As the market for low-carbon technologies expands, they will improve further in performance and price and become even more competitive. In Ted Turner's words, it is the greatest economic opportunity of the 21st century.

What the Paris climate conference can do is help accelerate the pace of technological adoption and change, toward the day when the cleanest energy sources are also the cheapest and thus become dominant. The payoff will be improved public health and increased economic well-being. Most importantly, the global climate will be prevented from going off the rails. The United Nations' Intergovernmental Panel on Climate Change (IPCC) was established by the World Meteorological Organization and the United Nations Environment Program in 1988 to review and assess published scientific, technical, and socio-economic reports on climate change, its potential impact, and options for adaptation and mitigation. Drawing on the voluntary engagement of hundreds of experts from leading academic and research institutions, the IPCC is the world's most authoritative scientific effort to understand and address changes in the Earth's climate.

To date, the IPCC has released five Assessment Reports, which have warned in increasingly strong and certain terms about the impacts of climate change on our global environment, society, and economy if serious action is not taken immediately. In 2014 the IPCC concluded in its Fifth Assessment Report: "Warming of the climate system is unequivocal, and since the 1950s, many of the observed changes are unprecedented over decades to millennia. The atmosphere and ocean have warmed, the amounts of snow and ice have diminished, and sea level has risen." With 95 per cent certainty, these scientists attribute the changes to human activity. The year 2014 ranks as Earth's warmest since modern recordkeeping began in 1880, according to two analyses by the US government and another by Japan. The world is already seeing and feeling the effects—in the form of rising sea levels and more extreme weather, heat waves, wildfires, and storms. These changes could lead to devastating effects on humanity and will only worsen if left unchecked.

The UN Foundation works to ensure that the science of climate change is communicated and understood globally. It also supports action on a range of global energy issues, especially Sustainable Energy for All, the initiative launched by UN Secretary-General Ban Ki-moon and co-led by World Bank President Jim Kim. Along with a goal of ensuring universal access to modern energy services, Sustainable Energy for All seeks to double the rate of improvement in energy efficiency globally and double the share of renewable energy in the global energy mix thus addressing climate change through critically needed changes to the world's energy systems within a framework of sustainable development.[22] The Paris Agreement marks the latest step in the evolution of the UN climate change regime, which originated in 1992 with the adoption of the Framework Convention. The UNFCCC established a long-term objective, general principles, common and differentiated commitments, and a basic governance structure, including an annual COP.

In the years since, the regime has evolved in different directions. The 1997 Kyoto Protocol took a more "top-down" but highly differentiated approach, establishing negotiated, binding emissions targets for developed countries, and no new commitments for developing countries. Because the United States did not join, and some countries that did set no targets beyond 2012, the protocol now covers less than 15 percent of global emissions. With the 2009 Copenhagen Accord and 2010 Cancun Agreements, parties established a parallel "bottom-up" framework, with countries undertaking national pledges for 2020 that represent political rather than legal commitments. This approach attracted much wider participation, including, for the first time, specific mitigation pledges by developing countries. However, countries' pledges fell far short of the reductions needed to meet the goal set in Copenhagen and Cancun of keeping average warming below 2°C above pre-industrial levels.

The negotiations toward a Paris agreement were launched with the Durban Platform for Enhanced Action adopted at COP 17 in 2011. The Durban Platform called for "a protocol, another legal instrument or an agreed outcome with legal force under the Convention applicable to all Parties," to apply from 2020, but provided no further substantive guidance.

COP 19 in Warsaw called on parties to submit "intended nationally determined contributions" (INDCs) well before the Paris conference, signalling an important bottom-up feature of the emerging agreement. Heading into Paris, more than 180 countries producing more than 90 percent of global emissions had submitted INDCs, a much broader response than many had anticipated.[23] The success of ongoing negotiations to establish a new global climate change agreement depends heavily on the agreement's acceptance by the world's major economies, including the United States. The new agreement is being negotiated under the United Nations Framework Convention on Climate Change (UNFCCC), a treaty with 195 parties that was ratified by the United States in 1992 with the advice and consent of the US Senate. US acceptance of the new agreement may or may not require legislative approval, depending on its specific contents.

The US law recognizes several routes for entering into international agreements. The most commonly known, under Article II of the Constitution, requires advice and consent by two-thirds of the Senate. In practice, however, the United States has accepted the vast majority of the international agreements to which it is a party through other procedures. These include congressional-executive agreements, which are approved by both houses of Congress, and presidential-executive agreements, which are approved solely by the president. The President would be on relatively firm legal ground accepting a new climate agreement with legal force, without submitting it to the Senate or Congress for approval, to the extent it is procedurally oriented, could be implemented on the basis of existing law, and is aimed at implementing or elaborating the UNFCCC. On the other hand, if the new agreement establishes legally binding emissions limits or new legally binding financial commitments, this would weigh in favour of seeking Senate or congressional approval. However, the exact scope of the President's legal authority to conclude international agreements is uncertain, and the President's decision will likely rest also on political and prudential considerations.[24]

The UN Security Council met on 20 July 2011 to debate the impact of climate change on maintaining international peace and security. The meeting was organized at the initiative of Germany during that country's presidency of the Council. In his introductory statement, UN Secretary-General Ban Ki-moon set the tone: climate change "not only exacerbates threats to international peace and security, it is a threat to international peace and security. Around the world, hundreds of millions of people are in danger of going short of food and water. That undermines the most essential foundations of local, national and global stability. Competition between communities and countries for scarce resources, especially water, is increasing, exacerbating old security dilemmas and creating new ones. Environmental refugees are reshaping the human geography of the planet, a trend that will only increase as desert advance, forests are felled and sea-levels rise. Mega-crises may well become the new normal." Effects of climate change on lives, infrastructure, institutions, and budgets he described as "an unholy brew that can create dangerous security vacuums."[25]

7.4 United Nations Development Programme (UNDP)

UNDP's vision is one where all developing countries can achieve a zero carbon and sustainable future. To achieve this vision, countries must rapidly transform their economies, such as by greening their energy, infrastructure, agriculture and other sectors. UNDP works with national, regional, and local planning bodies to help them respond effectively to climate change and promote low-emission, climate-resilient development. UNDP support focuses on three areas: connecting countries to knowledge, experience and resources to help people build a better life; helping countries build more resilient societies; and strengthening the capacity of countries to access, manage and account for climate finance.

Climate change and associated finance are at the heart of UNDP's strategic Plan, and our portfolio supports opportunities for countries to take action under the United Nations Framework Convention on Climate Change (UNFCCC) process. In partnership with the Global Environment Facility (GEF), Adaptation Fund (AF), as well as other multilateral and bilateral partners, UNDP has provided decades of strategic assistance in catalysing and leveraging climate investment into climate policies, technologies, practices and enterprises.[26]

Through UNDP's extensive work at global, national and regional levels, UNDP has learned that tackling climate change must be central to efforts to reduce poverty and sustain development. UNDP is working to help countries address the challenges of climate change and remains the largest service provider in the UN system on climate change adaptation and mitigation. UNDP supports countries to transition toward low-emission and climate-resilient sustainable development helps them to prepare for and build resilience to the impacts of climate change, and pursue low carbon development pathways that guarantee a cleaner, greener future. UNDP partners in many global dialogues, including the UNFCCC negotiations, as well as provides on-the-ground technical assistance for the many mechanisms on climate change. In this regard, UNDP has supported more than 140 countries to access and implement climate change initiatives.

- UNDP remains the largest service provider in the UN system supporting countries on climate change adaptation and mitigation.
- UNDP's support on adaptation.
- UNDP's support on mitigation.

7.5 United Nations Environment Programme (UNEP)

UNEP is one of three implementing agencies of the Global Environment Facility (GEF), along with the World Bank and UNDP. The GEF helps developing countries and those with economies in transition to meet the agreed incremental costs of measures designed to achieve global environmental benefits in six focal areas: biological diversity, climate change, international waters, ozone layer depletion, land degradation and persistent organic pollutants. UNEP provides support and the secretariat to the Scientific and Technical Advisory Panel (STAP) of the GEF, corporate support to the GEF—for example in the development of policy—and supports development and implementation of GEF-funded projects.[27]

For more than two decades UNEP has played a key role in United Nations efforts to address climate change and increase awareness among governments, the scientific and business communities and the general public. Assessments by the Intergovernmental Panel on Climate Change (IPCC)—established by UNEP and the World Meteorological Organization in 1988—underpinned negotiations for the Kyoto Protocol, and continue to provide the foundation for global decision making on climate change. The fourth IPCC assessment was released during 2007, with the three Working Groups of the IPCC each presenting their findings, culminating in a synthesis report and a summary for policymakers, released on 17 November 2007 in Valencia, Spain.[28]

The threat that climate change poses to peace, security and sustainable development led UN Secretary-General Ban Ki-moon to make it one of his top priorities on assuming office in January 2007. In September, Secretary-General Ban called a high-level meeting on climate change at UN headquarters in which UNEP participated. The meeting generated broad agreement among governments on the urgency of the climate challenge and that the principal arena for addressing it is the UN Framework Convention on Climate Change (UNFCCC). UNEP supports the UNFCCC process in a variety of ways, and will continue to support the Parties to the UNFCCC as they work to put in place a structure for reaching a post-2012 agreement that focuses on significantly reducing greenhouse gas (GHG) emissions and supporting for adaptation activities.[29]

7.5.1 UNEP-GEF and Climate Change

Examples of UNEP-GEF projects and activities are featured throughout this report. GEF climate change projects are organized into four areas: (i) removing barriers to energy efficiency and energy conservation; (ii) Promoting the adoption of renewable energy by removing barriers and reducing implementation costs; (iii) reducing the long-term costs of low greenhouse gas emitting energy technologies; and (iv) supporting the development of sustainable transport.[30]

7.6 Strategic Priorities of UNEP's Climate Change Programme

Within the Medium Term Strategy and under the climate change thematic priority, UNEP's objective is to strengthen the ability of countries to integrate climate change responses into national development processes. Consistent with the UNFCCC and within the broader United Nations approach for dealing with climate change, the objectives and expected accomplishments focus on providing environmental leadership in the four areas prominent in the international response to climate change: adaptation, mitigation, technology and finance, and their inter-linkages.

The UNEP expected accomplishments are as follows:

(i) That adaptation planning, financing and cost-effective preventative actions are increasingly incorporated into national development processes that are supported by scientific information, integrated climate impact assessments and local climate data.

(ii) That countries make sound policy, technology, and investment choices that lead to a reduction in greenhouse gas emissions and potential co-benefits, with a focus on clean and renewable energy sources, energy efficiency and energy conservation.

(iii) That improved technologies are deployed and obsolescent technologies phased out, financed through private and public sources, including the Clean Development Mechanism.

(iv) That increased carbon sequestration occurs through improved land use, reduced deforestation and reduced land degradation.

(v) That country policy-makers and negotiators, civil society and the private sector have access to relevant climate change science and information for decision-making.

The question today lies what kind of supportive or strategic role could or should UNDP, UNEP are playing in dealing with climate change and security change. The question regarding the role of UNEP and UNDP remains today is what kind of supportive or strategic role they could or should play in dealing with climate change and security change. The question is less and less one of whether climate change is a security threat multiplier, but one of how we can assess and manage the risk associated with climate change and its security implications as an international community. There can be little doubt today that climate change has potentially far-reaching implications for global stability and security in economic, social and environmental terms which will increasingly transcend the capacity of individual nation States to manage. In that context the sustainable development paths of individuals will increasingly be predicated upon the ability of the international community to act collectively in addressing these developments. Climate change will probably have many serious effects, particularly transition effects, on peoples and societies worldwide. The hardships of climate change are particularly likely to add to the burden of poverty and human insecurity of already vulnerable societies and weak governments.

Thus, climate change can be seen as a security issue in a broad sense, and efforts to halt or reverse it may well warrant a peace prize. However, so far there is little if any solid evidence that we are going to see an increase in armed conflicts as a result of climate change. Assuming such a link without the necessary evidence may lead peace-making astray and can eventually also undermine the credibility of the IPCC and the efforts to reach a consensus of knowledge about man-made climate change and a concerted global effort at mitigation and adaptation. The climate-conflict discourse is easily exploited by cynical governments and ruthless rebels who would like to evade any direct responsibility for atrocities and violence and prefer to put the blame on developed countries and their greenhouse gas emissions. Above all, blind reproduction of "folk wisdom" about the climate-conflict relationship will undermine serious research in this area and potentially leave as less well-prepared for dealing with the social effects of climate change once they really start kicking in.

7.6.1 The Bali Roadmap

The Bali Road Map was adopted at the 13th Conference of the Parties and the 3rd Meeting of the Parties in December 2007 in Bali. The Road Map is a set of forward-looking decisions that represent the work that needs to be done under various negotiating "tracks" at essential to reaching a secure climate future. The Bali Road Map includes the Bali Action Plan, which charts the course for a new negotiating process designed to tackle Climate Change. The Bali Action Plan is a comprehensive process to enable the full, effective and sustained implementation of the Convention through long-term cooperative action, now, up to and beyond 2012, in order to reach an agreed outcome and adopt a decision. All Parties to the Convention were involved in crafting the Bali Road Map.

7.6.2 The Copenhagen Accord

The 15th session of the Conference of the Parties to the UNFCCC and the 15th session of the Parties serving as the Meeting of the Parties to the Kyoto Protocol took place in Copenhagen, Denmark in 2009. The Copenhagen Accord contained several key elements on which there was strong convergence of the views of the Government's. This included the long-term goal of limiting the maximum global average temperature increase to more than 2°C above pre-industrial levels, subject to a review in 2015. It also included a reference to consider limiting the temperature increase to below 1.5 degrees -key demand made by vulnerable developing countries. Developed countries promised to provide US$30 billion for the period 2010-2012, and to mobilize long-term finance of a further US$100 billion a year by 2020 from a variety of sources. The work of the two central negotiating groups, the AWG-LCA and the AWG-KP was extended by the COP.

7.6.3 The Cancun Agreements

The Cancun Agreements, reached on December 11 in Cancun, Mexico, at the 2010 United Nations Climate Change Conference, represented key steps forward in capturing plans to reduce Greenhouse Gas emissions, and to help developing nations protect themselves from climate impacts and build their sustainable future. The main objectives include: (1) Mitigation; (2) Transparency of actions; (3) Technology; (4) Adaption; (5) Forests; (6) Capacity building; and (7) Finance.

The objectives also include setting up the Green Climate Fund to disburse $100 billion per year by 2020 to developing countries to assist them in mitigating Climate Change and adapting to its impacts.

7.6.4 The Durban Agreement

The United Nations Climate Change Conference at Durban in 2011, delivered a breakthrough on the international community's response to Climate Change. All Governments committed in Durban to a comprehensive plan that would come closer over time to delivering the ultimate objective of the Climate Change Convention: to stabilize Greenhouse Gas concentrations in the atmosphere at a level that will prevent our dangerous interference with the climate system and at the same time will preserve the right to sustainable development. The developing countries, especially the poorest and most vulnerable, will need much more support to adapt to the change that is already embedded in the global climate system.

The Durban outcomes looked to address these challenges in a more connected way by embodying a road map for implementation. On this map, four main areas of coordinator and complementary action and implementation, designed also to build and preserve trust among countries, were agreed viz. (i) Second commitment period of the Kyoto Protocol; (ii) The launch of a new platform of negotiations under the Convention to deliver a new and universal Greenhouse Gas reduction protocol, legal instrument or other outcome with legal force by 2015 for the period beyond 2020; (iii) Conclusion in 2012 of existing broad-based stream of negotiations; and (iv) To scope out and then conduct a fresh global Review of the emerging climate challenge, based on the best available science and data.

7.6.5 The Doha Climate Gateway

At the 2012 UN Climate Change Conference in Doha, Qatar (COP18/ CMP8), Government's consolidated the gains of the three years of International Climate Change negotiations and opened a gateway to necessary greater ambition and action on all levels. Among the many decisions taken, some were as follows:

- The governments strengthened their decision and set out a timetable to adopt a universal climate agreement by 2015, which will come into effect in 2020.
- They streamlined the negotiations, completing the work under the Bali Action Plan to concentrate on the new work towards a 2015 agreement under a single negotiating stream in the Ad hoc Working Group on the Durban Platform for Enhanced Action (ADP).
- They emphasized the need to increase their ambition to cut Greenhouse Gases (GHGs) and to help vulnerable countries to adapt.
- Launched a new commitment period under the Kyoto Protocol, thereby ensuring that this treaty's important legal and accounting models remain in place and underlining the principle that developed countries lead mandated action to cut Greenhouse Gas emissions.
- Made further progress towards establishing the financial and technology support and new institutions to enable clean energy investments and sustainable growth in developing countries.

The aim was that the world has a chance to stay below an agreed maximum 2°C temperature rise, to avoid serious Climate Change impacts. The governments agreed to find ways to scale up efforts before 2020 beyond the existing pledges to curb emissions.

7.7 Future of Paris Agreement and Climate Change and Role of India

In December 2015, nations of the world adopted the Paris Agreement to establishing an operational regime to stimulate climate action, after several failed attempts. But Paris Agreement became successful through changing the paradigm diplomacy. It focuses on a bottom-up structure for emissions targets based on nationally determined contributions and balanced by top-down provisions for achieving strong global emission goals and provisions of key accountability include reporting and review.[31]

Natalie Mahowald, a lead author of the UN report on climate change says, "Since the US has one of the biggest economies in the world and has contributed the most to climate change, it is incredibly important that the US return to the Paris agreement." He further said, "the US leadership and the US-China bilateral agreement to cut CO_2 emissions were keys to getting the Paris agreement on track". Therefore, continued involvement and leadership role of the US is a key in the effort to stopping climate change.[32] Both, China and the US are two largest economies in the world. Despite coercive and liberal approaches towards managing climate issues, some important objectives are also shared by the two such as recognizing the urgency of the challenge, develop new technologies which will constitute the foundations of the green economic future and both have been zeroed in on industrial policy to achieve their objectives of climate. Moreover, gaining the leadership of the global movement for mitigating climate change for both Xi and Biden remains a strategic mission because both Washington and Beijing understands well that the climate politics at the end is related to rearranging the global order.[33] Under the leadership of Xi Jinping, China's dynamic pursuit for "ecological Civilization" and newly elected president of the US, Joe Biden's assurance to making climate issues at the centre of US domestic and foreign policies are likely to change the prevailing conditions of the global discourse on environment. However, the new directions of Chinese and the US policies along with partnership with Europe and Japan will equally pressurize other states to re-organise their energy production and consumption, restructure economies, adopt new technologies and manage the internal political repercussion against the costs disseminated unequally among nations and within their class hierarchies.

As far as India's ability to influence the new geopolitics of climate change is concerned, largely depends on its domestic political resilience in adapting to the new imperatives. Because, when a democratic India still struggles to deal with the new internal conflicts centered on climate, China has already crafted a new template of "coercive environmentalism".[34] Earlier, China was the prime example of putting growth and development above considerations of ecological sustainability has changed in recent years. Now, China has decisively moved towards addressing many challenges of climate change. At present, China emits roughly 30 per cent of global GHGs but shutting down the coal-guzzling power plants shows it is reducing its emissions followed by the US around 15 per cent, Europe 10 per cent. India comes at the fourth position by flowing 7 per cent of the global total. But its per-capita emissions are less than half the global average. It is expected that US and China will stand together in the global negotiations. Delhi in this aspect, prepared better than earlier which was considered as problematic in addressing the climate change issues. But, now it has broken its nature of taking defensive stand by actively participating and shaping international debates on climate change. India is a strong advocate of global action on climate change as it is also has its civilisational links to the nature. Further, to achieve the 40 per cent of non-fossil-fuel installed electricity production capacity, India has committed to reducing its amount of GDP emissions by 33-35 per cent from 2005 levels by 2030. India's establishment of the International Solar Alliance with France can be seen as a major attempt towards it and became one of the few major climate stakeholders on track to over-achieve the targets it had committed.

Recently, the UN Secretary General has announced to seek a goal of net-zero emissions of GHGs by 2050 by considering the central objective of the UN in 2021 to establish a Global Coalition for Carbon Neutrality. India's engagement in global climate negotiations becomes important from the strategic perspective and informed by the building of a GREEN (Growth with Renewable Energy, Entrepreneurship and Nature) coalition of countries with development imperatives as well as climate perspectives.[35] Though, China's ecological activism has been considered as the model for "authoritarian environmentalism it is finding an echo among some of the Western environmentalists who believes planetary survival must be the foremost objective of the moment and all other considerations must be subordinated to it. But, whatever the merits of authoritarian environmentalism, it has little political chance of being replicated in democracies. Plural societies on the other hand, focused on improving the nature of liberal environmentalism that relies on political consensus in drafting new environmental norms and their effective enforcement along with the reliance on market-based mechanisms. However, Biden plans to enforce environmental regulations discarded by the Trump administration and enhancing the incentives for polluters to compensate for their violation of norms. For Biden, "Climate justice" is another important objective of domestic environmental policy. He is also looking beyond joining the Paris Accord to set accelerated timelines for reducing carbon emissions and give special emphasis on an early end to the worldwide use of coal.[36] As far as India's engagement with Biden by underlining climate change as an important area it has signaled its readiness to deal with a new phase in the global politics of environment. But diplomacy can only win time and space for India's internal adjustment to the new imperatives. Therefore, India's real test on climate change lies on building a new domestic consensus that can address the economic and political costs associated with an internal adjustment to the prospect of a great global rearrangement.[37]

NOTES

1. Stern N., The Economics of Climate Change: The Stern Review. Cambridge, UK: Cambridge University Press, 2007.
2. Nordhaus W.D., A Question of Balance: Weighing the Options on Global Warming Policies. New Haven: Yale University Press, 2008.
3. Bosetti V., C. Carraro, A. Sgobbi, and M. Tavoni, Modelling economic impacts of alternative international climate policy architectures: A quantitative and comparative assessment of architectures for agreement. In *Post-Kyoto International Climate Policy: Implementing Architectures for Agreement*, J.E. Aldy, R.N. Stavins, (eds.), Cambridge, UK: Cambridge University Press, pp.715–52, 2010.
4. Rayner S., How to eat an elephant: A bottom-up approach to climate policy, Climate Policy 10, 615–21, 2010.
5. Dubash N.K., Copenhagen: Climate of mistrust, Economic & Political Weekly 64, pp. 8–11, 2009. Accessed through: http://re.indiaenvironmentportal.org.in/files/COP.pdf on 25 November 2016.
6. Baer P., T. Athanasiou, S. Kartha, and E. Kemp-Benedict: Greenhouse development rights: A proposal for a fair global climate treaty, Ethics, Place & Environment 12, 267–81, 2009.
7. Okereke C., Climate justice and the international regime, Wiley Interdisciplinary Reviews: Climate Change 1,pp.462–74 2010.
8. Posner E., and C. Sunstein, Justice and climate change: The unpersuasive case for per capita allocations of emissions rights. In: Post-Kyoto International Climate Policy: Implementing Architectures for Agreement. J.E. Aldy, R.N. Stavins (eds.), Cambridge, UK: Cambridge University Press, pp.343–712010.
9. Somanathan E., What do we expect from an international climate agreement? A

perspective from a low-income country. In: Post-Kyoto International Climate Policy: Implementing Architectures for Agreement. J.E. Aldy, R.N. Stavins, (eds.), Cambridge, UK: Cambridge University Press, pp.599–617 2010.

10. Cao J. 2010c. Beyond Copenhagen: Reconciling International Fairness, Economic Development, and Climate Protection. The Harvard Project on International Climate Agreements. Accessed through: http://belfercenter.ksg.harvard.edu/files/CaoHPICADP44.pdfon 25 November 2016.
11. Sen A. The Idea of Justice. Cambridge M.A.: Belknap Press of Harvard University Press, 2009.
12. Cao J., Reconciling human development and climate protection: A multistage hybrid climate policy architecture. In: Post-Kyoto International Climate Policy: Implementing Architectures for Agreement. J.E. Aldy, R.N. Stavins, (eds.), Cambridge, UK: Cambridge University Press, pp.563–98 2010.
13. Bodansky D., Deconstructing the precautionary principle. In: Bringing New Law to Ocean Waters. D.D. Caron, H.N. Scheiber, (eds.), Leiden, Netherlands: Brill, 2004.
14. Wiener J.: Precaution. In: The Oxford Handbook of International Environmental Law. D. Bodansky, J. Brunnée, E. Hey, (eds.), New York: Oxford University Press, pp.597–612, 2007.
15. Urueña R., Risk and randomness in international legal argumentation, Leiden Journal of International Law 21. Chapter 12,pp787–822 2008.
16. Hammitt J., Stratospheric ozone depletion and global climate change. In: The Reality of Precaution: Comparing Risk Regulation in the United States and Europe. J. Wiener, M. Rogers, J. Hammitt, P. Sand, (eds.), Washington, DC: RFF Press, 2010.
17. IPCC: Climate Change 2007: Mitigation of Climate Change: Contribution of Working Group III to the Fourth Assessment Report of the Intergovernmental Panel on Climate Change. Cambridge, UK and New York: Cambridge University Press, 2007.
18. World Bank: World Development Report 2010: Development and Climate Change. Washington, DC: The International Bank for Reconstruction and Development, 2010.
19. Keohane, R.O., After Hegemony: Cooperation and Discord in the World Political Economy. Princeton, NJ: Princeton University Press, 1984.
20. Young, Oran R., International Cooperation: Building Regimes for Natural Resources and the Environment. Ithaca, NY: Cornell University Press, 1989.
21. Victor D.G. (2006) Toward Effective International Cooperation on Climate Change: Numbers, Interests and Institutions, Global Environmental Politics, 6:3, August 2006. Massachusetts Institute of Technology. Accessed through: http://www.uio.no/studier/emner/annet/sum/SUM4015/h08/Victor.pdf on 28November 2016.
22. Tackling the Global Climate Challenge, Accessed through http://www.unfoundation.org/what-we-do/issues/energy-and-climate/tackling-global-climate-challenge.html on 30 November 2016.
23. Outcomes of the U.N. Climate Change Conference in Paris 21st Session of the Conference of the Parties to the United Nations Framework Convention on Climate Change (COP 21) 12 November-December 2015, Accessed through https://www.c2es.org/international/negotiations/cop21-paris/summary on 27 November 2016.

24. Bodansky, Daniel, Legal Options for U.S. Acceptance of a New Climate Change Agreement, 'Sandra Day O'Connor College of Law, Arizona State University, P.5, May 2015.
25. The UN Environmental Programme in climate Change and International Security. Source: Population and Development Review, vol.37, No.3 September 2011, Population Council, Accessed through URL://www.jstor.org/stable/23036068 on 20 November 2016 pp.601-6.
26. UNDP and Climate Change Zero Carbon, Sustainable Development, Accessed through http://www.undp.org/content/dam/undp/library/Climate%20and%20 Disaster%20Resilience/UNDP_a nd_Climate_Change.pdfon 25 November, 2016, p. 6.
27. UNEP 2007Annual Report, Accessed through http://www.unep.org/PDF/ AnnualReport/2007/AnnualReport2007_en_web.pdfon 27 November 2016 p25.
28. Ibid, p.11.
29. Ibid.
30. Ibid.
31. Stern, Todd, 2018, The Paris Agreement and its future, Accessed at https:// www.brookings.edu/research/the-paris-agreement-and-its-future/ on 8 December 2020.
32. Newburger, Emma, 2020, Biden will rejoin the Paris Climate Accord. Here's what happens next. Accessed at https://www.cnbc.com/2020/11/20/biden-to-rejoin-paris-climate-accord-heres-what-happens-next-.html on 8 December 2020.
33. Mohan, C, Raja, 2020. Delhi's real test on climate change is on building political consensus to adapt to changed reality, The Indian Express, 8 December 2020. Accessed at https://indianexpress.com/article/opinion/columns/india-climate-change-global-warming-Indian Prime Minister-govt-7095665/ on 8 December 2020.
34. Ibid
35. Puri, Singh, Manjeev, 2020, India must forge coalition with like-minded nations on climate change, Accessed at https://indianexpress.com/article/opinion/columns/ climate-change-india-us-china-7094360/ on 9 December 2020.
36. Mohan, C, Raja, 2020. Delhi's real test on climate change is on building political consensus to adapt to changed reality, The Indian Express, 8 December 2020. Accessed at https://indianexpress.com/article/opinion/columns/india-climate-change-global-warming-Indian Prime Minister-govt-7095665/ on 8 December 2020.
37. Ibid.

8

Conclusion

Almost all countries missed the symbolic 9 February, 2020, deadline to strengthen plans to fight climate change under the Paris Agreement even though the United Nations says action in 2020 is vital to prevent runaway global warming. Only three nations have submitted upgraded climate plans nine months before the start of November's summit in Glasgow, adding to uncertainties after UK Prime Minister Boris Johnson sacked Claire O'Neill, who was due to preside at the talks. He has not yet named a replacement. The UN has billed the 2020 summit, known as Cop26, as a first test of a key principle in the 2015 Paris Agreement that all countries will ratchet up ambition every five years to do more to cut greenhouse gas emissions. Of almost 200 governments, only the Marshall Islands, Suriname and Norway, which issued a new plan on 7 February, have provided new NDCs to rein in their greenhouse gas emissions and adapt their economies to ever more heatwaves, droughts, more powerful storms and rising sea levels.

The main debate is on that how would changing climate patterns affect both inter-state relations and international and national security in a narrow, geo-strategic sense as well as the well-being and survival of human beings and humankind. On 11 September 2009, the UN Secretary General presented his report identified climate change as a "threat multiplier" that exacerbates existing threat, such as persistent poverty, week institutions for resource management and mistrust between communities. It identifies five further ways in which climate change might affect security: Climate change could threaten food security and human health and increase exposure to extreme events; it could undermine the stability of states by slowing or reversing development; it could increase the likelihood of domestic conflicts due to migration and depleting resources, with possible international ramifications; Disappearance of territory might bring about issues of sovereignty, rights and security; International conflict might be a result of climate change's impact or demarcated international resources.

The planet is under threat from anthropogenic activities human emissions, and the Paris climate deal is, at least, a first step toward fixing the problem. Ice sheets are starting to melt, coastlines are flooding from rising seas, and some types of extreme weather are growing worse. Yet some of the consequences of an overheated planet might be avoided, or at least slowed, if the climate deal succeeds in reducing emissions. But the Paris deal could represent the moment at which, because of a shift in global economic policy, the inexorable rise in planet-warming carbon emissions that started during the Industrial Revolution began to level out and eventually decline. At the same time, the deal could be viewed as a signal to global financial and energy markets, triggering a fundamental shift away from investment is coal, oil and gas primary energy sources toward zero -carbon energy sources like wind, solar and nuclear power.

The changes that led to the Paris accord came about through a mix of factors, particularly major shifts in the domestic politics and bilateral relationship of China and United States, the world's two largest greenhouse gas polluters. The Paris accord is a triumph for President Obama and Secretary of State John Kerry, who both lobbied hard for it, but it has outraged many Republicans who are skeptical of the extent of human-caused climate change and believe the deal favours environment ideology over economic reality. The environment has not typically played a major role in voters, choices, and the issue most likely overshadowed the election cycle by fears of terrorism, though the drought in California and severe weather in many parts of the country have raised concerns for many Americans.

With reference to climate change, there are stark differences of opinions between India and the US. The US tries to tackle climate change with the EPA, while India does the same with the National Action Plan for Climate Change (NAPCC).[1] The two countries are together responsible for an increased number of greenhouse gases emissions. Therefore, through their respective agencies, both countries have implemented policies to help combat climate change. However, it has been tough for both countries to understand the brevity of climate change. While the US has not ratified any treaty or agreement till the Paris Agreement, India has ratified both, the Kyoto Agreement, as well as the Paris Agreement.

While both the countries face almost similar threats with regard to climate change, they both have their own methods to try and tackle climate change. India and the US signed the Memorandum of Understanding on Energy Security, Clean Energy and Climate Change in November 2009. According to this MoU, both the countries agreed to cooperate on several issues like strengthening of R&D efforts for use of unconventional sources of cleaner fuels like coal bed methane and shale gas; better management of forest resources, reforestation and reducing carbon emissions resulting from deforestation and change in land use, and cooperation on adaptation measures by studying changing weather patterns.

Both the countries launched Mission Innovation at the 21st Conference of the parties (COP-21) to the United Nations Framework Convention on Climate change in Paris, pledging along with 18 other countries, to seek to double clean energy research and development investment over five years. The two countries announced a $30 million public-private research effort in small grid and grid storage.

On 31 August 2016, India and the United states held the second India-US Strategic and Commercial Dialogue in New Delhi. The United States must lead the world by working with China, Russia, India and the rest of the International Community to transform energy system away from fossil fuel towards energy efficiency and sustainable energy. We need a global commitment to reduce carbon pollution emission. The US needs to lead the international community in the light against climate change to maintain American economic strength and global security.

It was seen as a biggest opportunity for India and also for the new government to project India as a leader of the world at the Paris 2015. As William Antholis (2014) wrote that "Indian can move India to the front ranks of international leadership on climate change while also advancing its economic development. He notes that cooperation with the US can help him in that effort." As a developing nation and economy, sustainable energy will be in priority for economic growth in India. It is prominent for India to become a part of any solution of the climate crisis. India as a fourth largest greenhouse emitter, after China, US and EU, recognized that the climate change will have impact on India directly, through irregular ice melting in Himalayan, uncertainty in monsoon, declining rainfalls and rising sea levels. According to the recent World Health Organisation (WHO) global air pollution database, India has 14 out of the 15 most polluted cities in the world in terms of PM 2.5 concentrations and according to Central Pollution Control Board (CPCB), 31 and 195 cities exceeded the prescribed PM2.5 and PM10 limit, standard respectively. It is not easy for global community to decide the major causes, specially the big current and historical contribution of developed countries. India stands for the principle of Common but differentiated responsibilities (CBDR).

India and US, the world's oldest and largest democracies, together stand for the protections of global warming and climate change. At present, India and the US share an extensive and expanding cultural, strategic, military, and economic relationship which is in the phase of implementing confidence building measures (CBM) to overcome the legacy of trust deficit-brought about by adversarial US foreign policies and multiple instances of technology denial which have plagued the relationship over several decades. Unrealistic expectations after the conclusion of the 2008 US foreign policies and multiple instances of technology denial have plagued the relationship over several decades. India and the United States released a joint statement on June 7, 2016, at the White House and reviewed the bilateral ties between the two nations. The statement covers various issues pertaining to climate change, clean energy initiative, countering terrorism, etc.

The United States and India share common climate and clean partners in the fight against climate change. Furthermore, the two countries will pursue under the leadership of the G20 strong outcomes to promote improved heavy-duty vehicle standards and efficiency in accordance with their national priorities and capabilities. While the US works through the EPA (Environmental Protection Agency) to tackle climate change, India works through the machinery of its various ministries. Under the Obama administration, the Clean Air Act and the Clean Water Act galvanised the way states in the US tackle change. In India, the government highlighted eight important missions to nationally accomplish to tackle climate change. It can therefore be argued that the US, under the Obama administration and the Indian government worked in their respective ways and within their individual capacities to tackle climate change.

The US-India strategic partnership is being held upon to boost shared values to enable the two countries to work together on global issues and to support common strategic and security interests to increase mutual security against the common threats posed by intolerance, terrorism, and the spread of weapons of mass destruction. The successful transformation of the US-India relationship then can have a positive influence on the evolving future international system, which was against emphasized by President Obama in New Delhi. Between the two, there have been some differences, which included US concerns over India's Nuclear Weapons Programme and the direction of India's economic reforms. In the past, these concerns though dominated US thinking about India, but now the US perceives India as a growing world power with which it shares common strategic interests. As such a strong partnership between the two countries always remained a possibility to address differences and shape a dynamic and collaborative future, should a need arise in the future.

The India-US strategic partnership can be of great help given the many areas of convergence of political and strategic interests, with the United States having enduring vital strategic interests in the Middle East, Gulf region and south East Asia, connecting the world's largest reserves of the energy and the SE Asia providing the strategic choke points linking the Indian and Pacific Oceans, which are also of critical importance to India's national security interests. The US India partnership is based on shared values such as democracy, pluralism and rule of law. The two countries inked a ten-year defence framework agreement in 2005 to facilitate expanded bilateral security cooperation.

Recognizing the critical importance of increasing energy access, reducing greenhouse gas emissions, and improving resilience in the face of climate change, India and US started a new and enhanced strategic partnership on every security, clean energy, and climate change. Since 2014, they have made important strides together to advance this strategic partnership, and to our related cooperation on environmental stewardship. The United States and India recognize the urgency of the threat of climate change and are therefore committed to the key issues to support each other. Both the countries are strengthening and expanding the highly successful US-India Partnership to Advance Clean Energy (PACE) and achieve ambitious clean energy objectives and Nationally Determined contributions. After the Obama government, when it comes to climate change, many environmental campaigners are alarmed. This is understandable as Trump has sent mixed signals and we have much to learn about new administration's plan. But there is hope, and even a potential opportunity for India.

The primary aim of the current research was to carry out a detailed study of climate change and strategic cooperation between India and USA and its impact on international peace and security. The existing climate change literature shows that there is no common consensus on the climate change issue even though it is an imminent threat. Developed countries and developing countries are taking their own strategic stand on climate change; however, more and more common public is becoming aware of the change in the climate on Earth because of events like the bushfires in Australia and the floods in the United States in 2020.

The study has also examined the literature on climate change and its politics by developed states and role of India and the US for tackling climate change. By doing this, it has reached to a conclusion that each policy developed by any nation is dependent on the leader of the state as well. Up until, Obama was the president of the US, the policies towards climate change and the Paris agreement were being implemented at full speed. However, after the election of Donald Trump as the president of the US, things have taken a back seat.

Trump's role in evidently promoting domestic jobs and stimulate economic activities were considered as openly promoting the fossil-fuel industry that would cause increase in emissions. It reversing an order of 2015, mandating the US federal government agencies to reduce their own greenhouse gas emissions by 40 per cent in ten years, compared to 2008 levels. In finalising the Paris agreement, the US had played a pivotal role as it was not satisfied with the earlier international climate agreement hence never become member of it. So far, the Paris Agreement has also been very skilfully balanced. As a result, the implementation issues are not resolved yet. As far as the US' withdrawal from Paris Agreement is concerned, the trust between major economies damaged which was carefully built in two decades and created condition for gradually regaining the trust lying on foundations, needed to strengthen. Rather, Trump has weakened those existing fundamentals. In addition, his denial to climate science provided big reason to the other major polluters not to speed up their efforts.

Above all, he re-injected the denial of science into the public discourse on climate change. It is likely to change under Biden administration who aims to spend more than $2 trillion on energy and infrastructure for achieving the target net–zero emissions by 2050 and contain zero-emission power generation by 2035 at the domestic level, along with stronger regulation of polluters and focus on climate risks for companies and their supply chains. But, it has to demonstrate its climate leadership to the world through action rather than expressions.

At international level, Biden plans lies to re-engage with the Paris agreement as well as host a world summit to appeal participant countries for intensifying their climate pledge. In this context, India will not be an exception. It will put pressure on India to declare its plan for mid-century. The EU has already come up with a plan to introduce carbon border adjustment mechanism by 2021. Expectedly, the US can also introduce border tariff that both Republican and Democrats would agree has potential to create a climate related trade disputes. For India under such compulsions, has an opportunity to change the game and shift focus on climate actions rather than political rhetoric. It has to reach out to the Biden administration early and propose a bilateral deal drawing out four contours-changing narratives, facilitating investment, developing technology and increasing resilience. Currently, the narratives about climate goal is associated with increasingly terms for instance, net-zero, carbon-neutral and climate positive. Recently, China (2060), EU (2050) and Japan (2050) announced to achieve net-zero emissions are not suitable for the developing countries as it is too far in the nature and uncertain to have serious accountability and reduce growth opportunities for developing countries. It means lesser the action in the past or near future would tighten carbon budget in the future for them. Therefore, a changed narrative must ensure to place finest and proximate action and lessen the outlying promises. It is chance for India to powerfully demonstrate its climate actions to the world.

India has become first country in the world to have a cooling action plan and intends to seal 30 percent electric vehicles by 2030 in addition to 4,50,000 megawatts of renewable energy capacity target. Currently, hundreds of millions have electricity accessibility, clean cooking energy, and huge reduction in costs of LED light bulbs. On the Indian side, India has taken pro-active measures to tackle climate change under the leadership of current Prime Minister. Indian Prime Minister has been vocal about India's efforts to tackle climate change at the international stage and India has now become a leader in climate change conferences, leading the talks. Earlier, India never had such chances because it was considered as a developing nation, which would not give up using fossil fuels and non-renewable energy in order to become developed. However, current Indian Prime Minister has proven that even though India is developing, it is committed to reducing its greenhouse emissions and tackling climate change at a global level. The Indian government has introduced various schemes for electric vehicles, to reduce use of petrol, diesel vehicles. It has also introduced solar programmes across the country to generate electricity, showing that even though using renewable energy for development is a long and hard road, but India is not afraid to take it because it is committed to protect the environment.

It has also been observed that even though countries like the US and other developed countries are failing to tackle climate change, individuals are making a great difference at their levels. Greta Thunberg, a 14-year-old girl shook the world with her climate change appeal and has inspired millions of people to do best at their individual level to tackle climate change. The relations of climate change and security issues are also discussed under the proposed study explain the multifaceted relationship of countries. For example, While Amazon forest was burning in 2019; Brazil has not allowed any nation to intervene in the matter to douse the fires raging through the Amazon due to its security issues with neighbouring countries. However, other countries were also affected by this because the Amazon forests form the lungs of the earth and with their burning, the entire world would face severe climate change issues and hence, wanted to bring the fire under control, but Brazil, did not allow it. This in turn, created sour relations between many countries and Brazil prompting how important climate change is to security of nations.

Further, Climate change acts as an "Accelerant of Instability" or a "Threat Multiplier" that makes already existing problems worse. A Global Military Advisory Council on Climate Change Report in 2016 "Climate Change & Security in South Asia: Cooperating For Peace" by Lt. General Tariq Waseem Ghazi (Ret.) of Pakistan, Maj. General A.N.M. Muniruzzaman (Ret.) of Bangladesh, and Air Marshall A.K. Singh (Ret.) of India warns that unless measures are taken to address the threat of climate change, it could destabilize the South Asian region. In a region already affected by food and water insecurity, population displacement and weak disaster management infrastructure, you add climate change to the mix and it's kind of this classic threat multiplier. It takes all of these other problems and makes them worse. For example, Bangladesh is facing diverse challenges like economic development is intertwined with workers' rights and garment manufacturing; there is a growing threat of terrorism as Bangladeshi nationals return from war in Afghanistan; there are great power rivalries in the Bay of Bengal between China, India, and the United States. However, all of that is threatened by climate change. Rising sea levels threaten to make the homes of over 20 million people in Bangladesh uninhabitable. Changing ice melt from the Himalayas threaten the flow of water down the Brahmaputra and Ganges rivers, leading to greater floods in the monsoon season and drought in dry season. These changes will lead to migration, both internally and externally. Climate change will make all of the problems facing Bangladesh more difficult.

Problems like a lack of economic growth, endemic disease, hunger, and fresh water availability are a greater challenge to human security than simply a rise in temperature of 2 or 4 degree in average temperature. Terrorism, nuclear proliferation, and resource wars would seem to pose a greater threat to global security than just a difference of 2 degree in average temperature.

Unfortunately, climate change is what we call a 'ring road' issue; meaning that climate change affects all of these other threats. Unchecked, a warming of only 2°C will have significant impacts on water, food, and energy security. It will change disease vectors. It will drive migration. These changes, in turn, could affect state stability and harm global security.

There is no region of the world that faces more threats from climate change than South Asia. Of particular concern is the littoral surrounding the Bay of Bengal, including the Eastern Indian states of West Bengal and Odisha, Bangladesh, and coastal Burma. This region is uniquely vulnerable to a changing climate because of a combination of rising sea levels, changing weather patterns, and uncertain transboundary river flows. Away from the seashore, China holds the high ground in the Tibetan Plateau and the Himalayas, and complicates the geopolitical picture further by acting as the source of the region's fresh water. Climate change will not only affect American security through impacts on the economy and our physical infrastructure (roads, bridges, airports, etc.); it can also affect our domestic and international military bases. Physical changes to the environment such as flooding, drought and extreme weather events may disrupt US military capabilities and facilities, including military training ranges and bases. According to the US Department of Defense, there are a number of US military installations that are already at risk. The National Intelligence Council [NIC] judged that more than 30 US military installations were already facing elevated levels of risk from rising sea levels. DoD's operational readiness hinges on continued access to land, air, and sea training and test space.

Although sea-level flooding remains a major concern for the security of US military bases, a variety of other environmental threats pose serious risks as well. In September 2018, Marine Corps' Camp Lejeune was severely damaged by Hurricane Florence and subsequent flooding of the state. Estimates suggest that recovery will cost $3.6 billion. Only a month later in October 2018, Tyndall Air Force Base was hammered by Hurricane Michael. Previously home to many of the Air Force's F-22s, rebuilding will likely take years and cost upwards of $4.5 billion. The damage to these bases undermines the readiness of our fighting forces, diverting resources and attention away from the mission. In addition, military bases in the US are important for driving local economies and when they are destroyed by natural disasters, there is a ripple effect economically in the region.

According to a Department of Defense Report, the United States military manages property in all 50 states, 7 US territories and 40 foreign countries, comprising almost 300,000 individual buildings around the globe. These buildings are valued at $590 billion. The Army alone has over 14 million acres of property, 2000 installations and 12,000 historical structures. As the effects of climate change increase in many parts of the world, our investments and structures may be at risk of severe damage. Worryingly, recent studies have found that the military is not prepared to meet this threat.

The physical and operational security of US military bases around the world is a vital requirement to maintaining the national security of the United States and stability abroad. For example, the island of Diego Garcia in the Indian Ocean is a critical logistics hub for US and British forces in the Middle East and houses critical Air Force Satellite Control Network equipment. It is one of our most vulnerable military installations and one of the most important for our national security. Climate change is a threat to the United States and it must begin to confront those potential impacts on its military installations and missions around the world. The climate related events includes the recurrent flooding, drought, desertification, wildfires and thawing permafrost either caused damage or had the potential to damage 79 US military installations around the world.

It shows that despite Trump's denials, climate change has been increasingly weakening the national security of the United States. In contrast, Joe Biden, the newly elected president of the United States has committed to re-joining the Paris agreement on the day of his commencement to the office. He believes, climate change as the "number one issue facing humanity." It reflects in his presidential election campaign in which climate policy was positioned at the centre. He has affirmed a national transition from fossil fuels to renewable energy which would provide a boost to the global efforts for turn down the climate crisis and mobilise financial support for low-carbon transition in countries like India.

Further, initiating transition of the American energy economy away from oil to renewable and an emissions–free power sectors by 2035, also remains part of it. For his climate plan, promised to spend $2 trillion (around INR 150 trillion). It includes bringing back and strengthening the environmental safe guards abandoned by Trump, upgrading millions of buildings to the highest possible energy–efficiency levels, manufacturing clean vehicles and setting up an office of environmental and climate justice. But, Biden administration's climate plans have to go through the Republican–controlled senate and a conservative US Supreme Court will be major obstacle. Reasonably, Republican senators are broadly, against any kind of climate change legislation or energy reforms. Meanwhile, conservative justice army Convey Barrett's categorisation of the climate crises as a "very contentious matter of public debate" during a hearing on her Supreme Court nomination, reinforced the fears and court may block the Democrat administration's plan on strong climate action.

In the fight against climate change, the US matters the most. Along with China, it is the biggest emitter of greenhouse gas (GHG) emissions in the world. However, China recently announced that it plans to become carbon neutral by 2060. The Berlin–based institute, Climate Analytics' Climate Action Tracker (CAT), stresses that, if both China and US can achieve the target of becoming Carbon neutral, would be adequate to limit global heating and get back to 2.3-2.4°C above pre-industrial levels by 2100. Therefore, the 1.5-2°C aim becomes more realistic with a US that honours the Paris agreement.

The Paris agreements seeks to keep the rise in global temperatures within 2°C, associated with pre-industrial era, that cannot possibly be achieved without active participation of the US. Besides, the US still remains the second largest emitter of greenhouse gases in the world after china. Barack Obama under his administration, promised to cut its greenhouse gas emissions by 26-28 percent by 2030. In comparison of 2005 as standard time as per his commitment to the Paris Agreement, the US emissions were at its peak during 2005 with more than seven billion tonnes of carbon dioxide equivalent has been emitted. Since then, in the lack of clearly defined deadline, the process of emissions has reached to lower level. In 2018, the US had emitted more than 6.6 trillion tonnes of carbon dioxide–equivalent. The Paris agreement target, anticipated that the US would reduce its emissions in coming decades by at least 1.5 billion tonnes and increase afterwards. But, its withdrawal from Paris agreement, it has no obligation to achieve this target. It has serious implications on the ability of the Paris agreement to meet its objective. Hundreds of billions of dollars required each year for climate finance to enable the transitions towards a low carbon economy. At this point, the absence of the US as a key facilitator from this process has been a huge setback. The US ability to mobilise climate funds, particularly from private corporations, which is key to achieving the 2 °C target is more crucial.

However, under Biden leadership, US expected to experience policy reversals on climate change and re-joining the Paris agreement is one of them. But, Biden is likely to face more challenges, particularly trying to re-establishing trust in the US for its climate actions. In 2014, the leaders of two countries, the US president Barack Obama and Chinese president Xi Jinping reached to an agreement of Climate Change, seen as largest carbon polluter historically and currently came to act collectively. But the deal had disturbed the unity of developing country and India in particular. It became trapped in a climate formation and encircled by greater economies demanding it for more climate action. Such demands arise due to India's highest level of emitting, despite being poor. But, under the new US presidentship, Joe Biden, provides strategic opportunities for India through historic Biden-

Indian Prime Minister deal in 2021. Biden's determination boosts the prospect of renewed efforts against the climate crisis.

As Biden is not willing to have second term, its climate legacy compels him to act strongly. India can become a close ally with the US and worked together, would help to pressurise China for submitting details about its plan to 2060 or cancel its funding of thermal power assets in the belt and road initiative countries. Also, they could ratify the Kigali amendment to the Montreal Protocol meant to cut down the production and consumption of Hydrofluorocarbons. For institutional investment, India has a huge demand in domestic as well as foreign for its low–carbon growth. The both countries can also cooperate to bridging the demand-supply gap, lower risk perceptions; de-risk investments, and promoting investment relationship at state level help to democratise and decentralise low carbon pathways. India's Science, Technology and Innovation Policy 2020 have clear focus on low carbon technologies. Thus, two countries can leverage their priorities in the area of R&D. The two significant areas where both can collaborate are green hydrogen and carbon dioxide removal. The alliance of India and US in a Global Green Hydrogen Alliance (GGHA) would draw global attention towards the technology transforming industrial production processes.

Significantly in 2020, both countries have faced remarkable events due to extreme weather such as cyclones and forest fires triggered billions of dollars to overcome the situation. To effectively resolve such issues, India and US could initiate multilateral approach based on the chronic risks. It would signify their stronger cooperation through India's sponsored Coalition for Disaster Resilient Infrastructure, developing financial solutions, to upgrade older infrastructure and come up with new insurance scheme for climate risks. For India, it's a time to grab the opportunity and act independently instead of react under pressure. Currently, the US has massive interest and expectation regarding climate action and India's potential is already proved in this context. Such interests would unite the two countries as well as strengthen their ties beginning with Biden-Indian Prime Minister climate greeting. At the recent G-20 meeting, Indian Prime Minister said, the climate change must be fought in an integrated, comprehensive and holistic way. He further emphasised, the entire world can progress faster if there is greater support of technology and finance to developing nations. Also at the G20 summit side line event 'Safeguarding the Planet: the Circular Carbon Economy Approach', the Prime Minister said, "India is not only meeting its Paris Agreement targets, but also exceeding them."

By mentioning India's traditional ethos, he said "India's adoption of low-carbon and climate resilient development practices are inspired by our traditional ethos of living in harmony with the environment and the commitment of my government". He claimed, "For humanity to prosper, every single individual must prosper. Rather than seeing labour as a factor of production alone, the focus must be on human dignity of every worker". According to him, such approach would be the best guarantee for safeguarding the planet. India is a signatory to 2015, the Paris Agreement under the United Nations Framework Convention on Climate Change. By taking it into consideration, India has applied concrete action in various areas under which making of LED lights popular that saves 38 million tonnes of carbon dioxide emission annually, asserted by Mr. Indian Prime Minister. He also proclaimed, "We will meet our new goal of 175 gigawatts of renewable energy well before the target of 2022. Now, we are taking a big step ahead by seeking to achieve 450 gigawatts by 2030." Ultimately, the climate century is upon us: the earth is warming, humans are to blame, and we must take immediate action now to prepare for climate change's massively disruptive consequences. Indeed, both the congressionally-mandated 2018 National Climate Assessment (NCA) and United Nations Intergovernmental Panel on Climate Change reports make clear that the window to take collective action to reduce worldwide greenhouse gas emissions is shrinking. And advances in the field of climate attribution science demonstrate that climate change plays a major role in the frequency and intensity of extreme weather events. No longer can climate change be categorized solely as an environmental issue. It is a grave threat to national security. Indeed, it may be the threat. In a cruel irony, the poorest nations of the world that contributed the least to global warming will bear the brunt of climate change's impacts. What we, as a society, choose to do (or not do) now will define the health and welfare of future generations. Their fate is increasingly shaped by climate change's dramatic, erratic, and catastrophic national security effects.

NOTES

1. Comparative Analysis of developing country national climate change strategies: Accessed through http://www.usclimatenetwork.org/resource-database/WRI%20Comparative%20Analysis%20OF%20Developing%20Country%20National%20 Climate%20 Change%20Strategies.pdf on 12 February 2017.

APPENDIX

Important Points of Climate Change Conferences and UN Framework Convention on Climate Change

The United Nations Framework Convention on Climate Change (Convention) is the guiding agreement that governs most multilateral negotiations to address the threat of climate change. However, while many commitments to reduce emissions have been made under the Convention, emissions continue to rise, causing greenhouse gases to accumulate in the atmosphere.[1]

The IPCC's first report in 1990 fed directly into the 1992 signing of the Convention at the Earth Summit in Rio de Janeiro, Brazil. The ultimate objective of the Convention is the "Stabilization of greenhouse gas concentrations in the atmosphere at a level that would prevent dangerous anthropogenic interference with the climate system. Such a level should be achieved within a time-frame sufficient to allow ecosystems to adapt naturally to climate change, to ensure that food production is not threatened and to enable economic development to proceed in a sustainable manner."[2]

The Conference of the Parties (COP) is the main decision-making organization (the "supreme body") of the UNFCCC and meets once a year to review the implementation of the Convention and make decisions necessary to promote its effectiveness. As the Convention itself is a "framework," one important aspect of the COP is to decide whether to adopt additional specific legal instruments or agreements that might further the Convention's objective.[3]

The convention was produced at the 1992 UN Conference on Environment and Development with the stated aim of reducing global greenhouse gas emissions to a level that would prevent dangerous anthropogenic (human induced) interference with the climate system. Signatories agreed to take account of climate change in their domestic policy and to develop national programmes that would slow its progress. However, no mandatory targets were established for the reduction of emissions so the treaty remained legally non-binding. Instead it operates as a 'framework' document, with provisions for regular updates and amendments.

The first of these additions was the Kyoto Protocol in 1997. Under the protocol, 37 developed countries are committed to reducing their collective emissions of six greenhouse gases to at least 5 per cent below 1990 levels.

The members of the UNFCCC meet on an annual basis. The conference in Indonesia in 2007 led to the creation of the Bali Roadmap, which timetables negotiations for a protocol to succeed Kyoto, a process continued at the 2008 conference in Poland. The subsequent Copenhagen Accord of 2009 was not legally binding and failed to set out concrete measures for tackling climate change. A package of decisions to encourage all governments to work towards a low-emissions future was adopted at the 16th conference held in 2010 in Cancun, Mexico. The 2011 conference, held in Durban, South Africa, advanced negotiations on the implementation of the Kyoto Protocol, the Bali Action Plan and the Cancun Agreements. At the 18th conference in Doha, Qatar in 2012 plans were laid for the development of a successor protocol by 2015 to be implemented by 2020.

This entered into force on 21 March 1994. Today, it has near-universal membership. The 197 countries that have ratified the Convention are called Parties to the Convention. In 1992, through the Convention, developed countries agreed to voluntarily reduce emissions individually or jointly to 1990 levels by the year 2000. This was Canada's domestic target at the time. Developed countries also pledged to provide "new and additional" financial resources to help developing countries establish greenhouse gas emissions inventories and finance their efforts in meeting the substantial commitments that all parties including developing countries, agreed to in article 4, paragraph 1, of the Convention ("Commitments").[4]

Because developed countries have, historically, produced the largest share of global emissions, and since they have greater technological and economic capacity than developing countries to initiate and manage change, it was accepted in the Convention that they have a responsibility to lead the global effort to reduce emissions. This is one interpretation of "common but differentiated responsibilities," a key concept in the Convention.[5]

Despite the voluntary pledges made under the Convention, greenhouse gas emissions in many developed countries continued to rise. As a result, a number of nations approved an addition to the treaty and, in 1997, the Kyoto Protocol was finalized under the Convention. The Kyoto Protocol has more powerful measures than the Convention, some of which are legally binding.[6] Both the initial voluntary targets of the Convention and the binding targets of the Kyoto Protocol were seen as preliminary steps to further significant global action. UNFCCC negotiations are now taking place within the context of having accepted the significant challenge of avoiding a 2°C threshold.[7]

Global politics is also an important context for the negotiations. As a result of the poor record of developed countries in general, developing countries doubt the commitment of developed countries to lead the effort to reduce global emissions. Developing countries also believe that developed countries have not helped them sufficiently to finance sustainable development practices and adapt to climate change, although some recent decisions at the UNFCCC may help address these issues.[8]

Meanwhile, the urgency to reduce emissions to meet the Convention's objective mounts. Models of the climate system suggest that, even if greenhouse gas emissions ceased to increase, temperatures would remain elevated for centuries. Also, current fossil fuel infrastructure investments will lock in associated future emissions for decades. Meanwhile, negotiations and implementation of an agreement must overcome the significant political dynamic between developed and developing countries that has prevented advancement toward an effective agreement.[9]

Although an increase in average global temperature of 4°C may constitute a more certain danger, the UNFCCC has accepted a threshold of 2°C as defining a danger to the climate system. As a result of the difficulties in reducing global emissions, it is becoming likely that this threshold will be passed. However, despite the failure of the UNFCCC to come to an effective international agreement, it has focused attention on climate change, catalysing domestic actions and international activities that were unlikely to have occurred in its absence. Recent data suggest that emissions in 2012 increased at a lower rate than in previous years. With the hope that current actions, particularly fuel switching, may be lowering the rate at which emissions are increasing, much more will have to be done to actually decrease emissions in order to lower atmospheric greenhouse gas concentrations sufficiently to meet the targets of the UNFCCC.[10]

Kyoto Protocol

The Kyoto Protocol is an international agreement linked to the United Nations Framework Convention on Climate Change. The major feature of the Kyoto Protocol is that it sets binding targets for 37 industrialized countries and the European community for reducing greenhouse gas (GHG) emissions. These amount to an average of five per cent against 1990 levels over the five-year period 2008-12.

The treaty also established an international trading system, which allows countries to earn credits towards their emission target by investing in emission clean-ups outside their own country. According to the Global Carbon Atlas, the largest contributors of greenhouse gases in 2014 were China and the United States.

The Rio Convention, also known as the "Earth Summit" produced the Rio Declaration and Agenda 21 (a plan of action for the UN organizations, Governments, and Major Groups in areas where human activities have a negative impact on the environment) and also led to agreement on two other conventions which became open for signature the Framework Convention on Climate Change (FCCC) and Biological Diversity, both legally binding.[11]

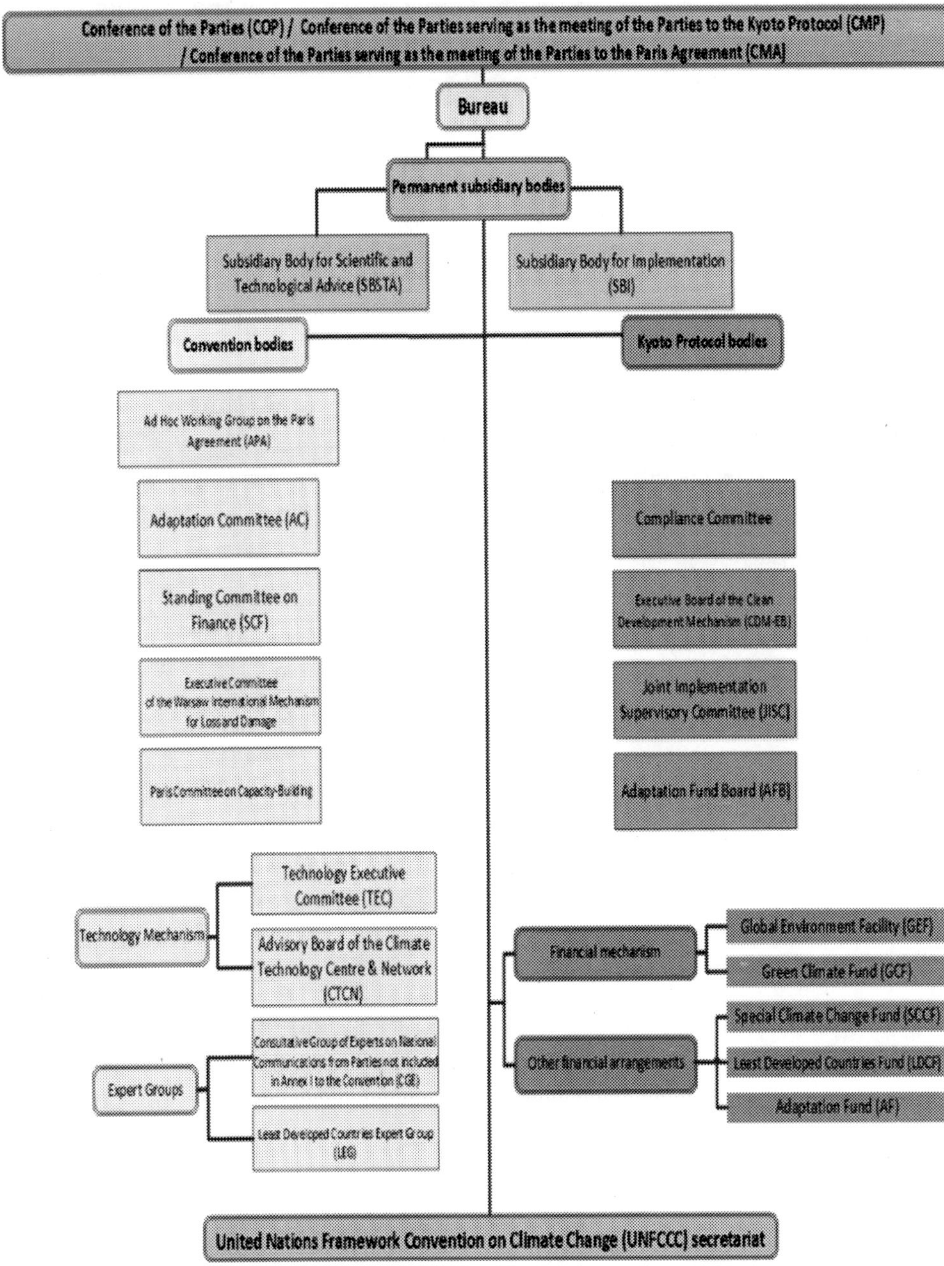
Conference of the Parties (COP) / Conference of the Parties serving as the meeting of the Parties to the Kyoto Protocol (CMP) / Conference of the Parties serving as the meeting of the Parties to the Paris Agreement (CMA)
Bureau
Permanent subsidiary bodies
Subsidiary Body for Scientific and Technological Advice (SBSTA)
Subsidiary Body for Implementation (SBI)
Convention bodies
Kyoto Protocol bodies
Ad Hoc Working Group on the Paris Agreement (APA)
Adaptation Committee (AC)
Standing Committee on Finance (SCF)
Executive Committee of the Warsaw International Mechanism for Loss and Damage
Paris Committee on Capacity-Building
Compliance Committee
Executive Board of the Clean Development Mechanism (CDM-EB)
Joint Implementation Supervisory Committee (JISC)
Adaptation Fund Board (AFB)
Technology Mechanism
Technology Executive Committee (TEC)
Advisory Board of the Climate Technology Centre & Network (CTCN)
Expert Groups
Consultative Group of Experts on National Communications from Parties not included in Annex I to the Convention (CGE)
Least Developed Countries Expert Group (LEG)
Financial mechanism
Global Environment Facility (GEF)
Green Climate Fund (GCF)
Other financial arrangements
Special Climate Change Fund (SCCF)
Least Developed Countries Fund (LDCF)
Adaptation Fund (AF)
United Nations Framework Convention on Climate Change (UNFCCC) secretariat

The FCCC can be considered the most important of the products of the Rio Convention in terms of leading to the preparation of the Kyoto Protocol. The FCCC was signed by 154 countries in 1992. Its key points were:

1) Stabilization of the amount of greenhouse gases in the atmosphere, while also ensuring food production is not put under threat, and allowing for economic development to proceed.
2) Developed countries should take the initiative in reducing levels of greenhouse gases in the atmosphere.
3) No specific greenhouse-gas-reduction aims, time frames, or penalties for violators were agreed on.
4) The participating countries decided on meeting at COPs (Conferences of the Parties) on a regular basis to work on the implementation of the Convention's objectives.

After meeting for 2 COPs in Berlin and Geneva, eventually on the third COP held in Kyoto, Japan in 1997 the member countries prepared the Kyoto Protocol.[12]

The Kyoto Protocol decided upon:

1) emission-reduction targets of greenhouse gases for each of the member countries;
2) the greenhouse gas emission-trading program; and
3) holding future meetings to set penalties for violators of the established targets and regulation rules of the emission-trading program.

In 2001, George W. Bush was elected president of the USA. Unlike his predecessor, President Clinton, President Bush did not support the Protocol and announced that the United States withdraws the possibility of ratifying it. This act put into question the whole concept of addressing the issues of climate change on a global level. Furthermore, in order for the Protocol to be put into action it had to be ratified by at least 55 nations of the UN Framework Convention, with developed countries representing a total of 55 per cent of the greenhouse gas emissions in 1990. With US holding a 36 per cent portion of this share, a ratification seemed a difficult task.[13]

The Kyoto Protocol finally came into effect on February 16, 2005.

The greenhouse effect is the warming effect of the sun on greenhouse gases, such as carbon dioxide, that act to trap this heat in our atmosphere. The more these gases exist, the more heat is prevented from escaping into space and, consequently, the more the earth heats up.

Although the greenhouse effect is necessary for survival on earth, an overabundance of greenhouse gas emissions increases global warming beyond what is desirable. The purpose of the Kyoto Protocol is to stabilize human-generated emissions at a level that will not inflict further harm on the atmosphere.

It took all of one year for the member countries of the Framework Convention on Climate Change to decide that the Convention had to be augmented by an agreement with stricter demands for reducing greenhouse-gas emissions. The Convention took effect in 1994, and by 1995 governments had begun negotiations on a protocol—an international agreement linked to the existing treaty, but standing on its own. The text of the Kyoto Protocol was adopted unanimously in 1997; it entered into force on 16 February 2005.[14]

The Protocol's major feature is that it has mandatory targets on greenhouse-gas emissions for the world's leading economies which have accepted it. These targets range from –8 per cent to +10 per cent of the countries' individual 1990 emissions levels "with a view to reducing their overall emissions of such gases by at least 5 per cent below existing 1990 levels in the commitment period 2008 to 2012." In almost all cases-even those set at +10 per cent of 1990 levels-the limits call for significant reductions in currently projected emissions. Future mandatory targets are expected to be established for "commitment periods" after 2012. These are to be negotiated well in advance of the periods concerned.[15]

The Kyoto Protocol was adopted as the first addition to the United Nations Framework Convention on Climate Change (UNFCCC), an international treaty that committed its signatories to develop national programs to reduce their emissions of greenhouse gases. Greenhouse gases, such as carbon dioxide (CO_2), methane (CH_4), nitrous oxide (N_2O), per fluorocarbons (PFCs), hydro fluorocarbons (HFCs), and sulphur hexafluoride (SF_6), affect the energy balance of the global atmosphere in ways expected to lead to an overall increase in global average temperature, known as global warming. The Kyoto Protocol committed most of the Annex I signatories to the UNFCCC (consisting of members of the Organisation for Economic Co-operation and Development and several countries with "economies in transition") to mandatory emission-reduction targets, which varied depending on the unique circumstances of each country. Other signatories to the UNFCCC and the protocol, consisting mostly of developing countries, were not required to restrict their emissions. The protocol entered into force in February 2005, 90 days after being ratified by at least 55 Annex I signatories that together accounted for at least 55 percent of total carbon dioxide emissions in 1990.

The protocol provided several means for countries to reach their targets. One approach was to make use of natural processes, called "sinks," that remove greenhouse gases from the atmosphere. The planting of trees, which take up carbon dioxide from the air, would be an example. Another approach was the international program called the Clean Development Mechanism (CDM), which encouraged developed countries to invest in technology and infrastructure in less-developed countries, where there were often significant opportunities to reduce emissions. Under the CDM, the investing country could claim the effective reduction in emissions as a credit toward meeting its obligations under the protocol. An example would be an investment in a clean-burning natural gas power plant to replace a proposed coal-fired plant. A third approach was emissions trading, which allowed participating countries to buy and sell emissions rights and thereby placed an economic value on greenhouse gas emissions. European countries initiated an emissions-trading market as a mechanism to work toward meeting their commitments under the Kyoto Protocol. Countries that failed to meet their emissions targets would be required to make up the difference between their targeted and actual emissions, plus a penalty amount of 30 percent, in the subsequent commitment period, beginning in 2012; they would also be prevented from engaging in emissions trading until they were judged to be in compliance with the protocol. The emission targets for commitment periods after 2012 were to be established in future protocols.

Challenges

Although the Kyoto Protocol represented a landmark diplomatic accomplishment, its success was far from assured. Indeed, reports issued in the first two years after the treaty took effect indicated that most participants would fail to meet their emission targets. Even if the targets were met, however, the ultimate benefit to the environment would not be significant, according to some critics, since China, the world's leading emitter of greenhouse gases, and the United States, the world's second largest emitter, were not bound by the protocol (China because of its status as a developing country and the United States because it had not ratified the protocol). Other critics claimed that the emission reductions called for in the protocol were too modest to make a detectable difference in global temperatures in the subsequent several decades, even if fully achieved with US participation. Meanwhile, some developing countries argued that improving adaptation to climate variability and change was just as important as reducing greenhouse gas emissions.

The Kyoto Protocol is a complicated agreement that has been slow in coming there are reasons for this. The Protocol not only has to be an effective against a complicated worldwide problem it also has to be politically acceptable. As a result, panels and committees have multiplied to monitor and referee its various programmes, and even after the agreement was approved in 1997, further negotiations were deemed necessary to hammer out instructions on how to "operate" it. These rules, adopted in 2001, are called the "Marrakesh Accords."[16]

- There is a delicate balance to international treaties. Those appealing enough to gain widespread support often aren't strong enough to solve the problems they focus on. (Because the Framework Convention was judged to have this weakness, despite its many valuable provisions, the Protocol was created to supplement it.) Yet treaties with real "teeth" may have difficulty attracting enough widespread support to be effective.[17]
- Some mechanisms of the Protocol had enough support that they were set up in advance of the Protocol's entry into force. The Clean Development Mechanism, for example-through which industrialized countries can partly meet their binding emissions targets through "credits" earned by sponsoring greenhouse-gas-

reducing projects in developing countries—already had an executive board before the Kyoto Protocol entered into force on 16 February 2005.[18]

At the 18th Conference of the Parties (COP18), held in Doha, Qatar, in 2012, delegates agreed to extend the Kyoto Protocol until 2020. They also reaffirmed their pledge from COP17, which had been held in Durban, South Africa, in 2011, to create a new, comprehensive, legally binding climate treaty by 2015 that would require greenhouse-gas-producing countries—including major carbon emitters not abiding by the Kyoto Protocol (such as China, India, and the United States)—to limit and reduce their emissions of carbon dioxide and other greenhouse gases. The new treaty, planned for implementation in 2020, would fully replace the Kyoto Protocol.

Timeline

December 1-11, 1997: The Conference of the Parties to the United Nations Framework Convention on Climate Change is held in Kyoto, Japan. Over 150 nations attend and adopt the first international treaty on controlling and reducing greenhouse gases.

November 2, 1998: In Buenos Aires, Argentina, 160 nations meet to work out details of the protocol and create the "Buenos Aires Action Plan."

March 27, 2001: EPA administrator Christine Todd Whitman states that the Bush administration has "no interest in implementing" the Kyoto Protocol.

July 23, 2001: Negotiators from 178 countries meet in Bonn, Germany, and agree to adopt the protocol. The United States doesn't participate.

November 10, 2001: Representatives from 160 countries meet in Marrakech, Morocco, to further work out details of the protocol.

November 18, 2004 : The Russian Federation ratifies the protocol, giving new hope that it can be implemented, even without the United States.

February 16, 2005: The Kyoto Protocol takes effect.

December 12, 2011: Canada officially renounces the Kyoto Protocol. Environment Minister Peter Kent says Kyoto's goals are unworkable because the US and China never agreed to Kyoto, and that a new pact is needed to address emissions.

December 2012: The Kyoto Protocol is extended to 2020 during a conference in Doha, Qatar.

June 23, 2013: Afghanistan adopts the Kyoto Protocol, becoming the 192nd participant.[19]

The Bali Road Map

At the 13th Council of the Parties to the UN Framework Convention on Climate Change, delegates negotiated for two weeks about negotiations for the next two years on how to address climate change starting two years after their deadline to agree. The result of their effort is a series of decisions that, taken altogether, is called the Bali Roadmap. This plan consisted of two tracks, one under the Kyoto Protocol and the other under the Convention. In addition to the AWG-KP, the UNFCCC set up a separate, ad hoc working group to recommend ways to enable "the full, effective and sustained implementation of the Convention through long-term cooperative action (AWG-LCA). Only parties to the Kyoto Protocol could take part in the AWG-KP, while all parties to the Convention could take part in the AWG-LCA."

The Bali Road Map was adopted at the 13th Conference of the Parties and the 3rd Meeting of the Parties in December 2007 in Bali. The Road Map is a set of forward-looking decisions that represent the work that needs to be done under various negotiating "tracks" that is essential to reaching a secure climate future.[20]

The Bali Road Map includes the Bali Action Plan, which charts the course for a new negotiating process designed to tackle climate change. The Bali Action Plan is a comprehensive process to enable the full, effective and sustained implementation of the Convention through long-term cooperative action, now, up to and beyond 2012, in order to reach an agreed outcome and adopt a decision. All Parties to the Convention were involved in crafting the Bali Road Map. The COP decided that the process would be conducted under a subsidiary body under the Convention, the Ad Hoc Working Group on Long-term Cooperative Action (AWG-LCA). The Bali Action Plan is divided into five main categories: shared vision, mitigation, adaptation, technology and financing. The shared vision refers to a long-term vision for action on climate change, including a long-term goal for emission reductions. The AWG-LCA subsequently split the work streams into components under those five parts.[21]

The Bali Action Plan is divided into five main categories: shared vision, mitigation, adaptation, technology and financing. The shared vision refers to a long-term vision for action on climate change, including a long-term goal for emission reductions. The AWG-LCA subsequently split the work streams into components under those five parts.[22]

Working in parallel would be the Ad Hoc Working Group on Further Commitments for annex I Parties under the Kyoto Protocol (AWG-KP) negotiations, which was established in Montreal in 2005. Until COP17 in Durban, the main focus of the negotiations under the KP had been to decide what to do when its first commitment period expired in 2012. A decision was reached in Durban to move into a second commitment period in 2013, with Annex I parties submitting their quantified emission reduction targets in May 2012, to be adopted at COP18 in Qatar in December 2012.[23]

The Road to Agreement

The Bali Action Plan was highly ambitious. In terms of the time lines it spelled out, it may have been overly optimistic, and underestimated the complexity both of climate change as a problem and of crafting a global response to it.

At COP15 in Copenhagen, the Conference of the Parties extended the AWG-LCA's mandate, enabling it to continue its work with the aim of presenting the outcome of this work at COP16 in Cancun in 2010. COP15 advanced many key issues.

- It raised climate change policy to the highest political level;
- It advanced the negotiations on the infrastructure needed for well-functioning, global climate change cooperation;
- It produced the Copenhagen Accord. It was not adopted by all governments, but it advanced a number of key issues; and

- It committed developed countries to $30 billion fast-start financing (in 2010-2012) for adaptation and mitigation in developing countries, with priority given to the least developed countries.

All of this momentum was built upon in Cancun in 2010, when governments drew up the Cancun Agreements, a set of significant decisions to respond to the long-term challenge of climate change collectively and comprehensively, now and over time. Countries decided to make their emission reduction pledges official, in what was the largest collective effort the world has ever seen to reduce emissions in a mutually accountable way. Click here for more information on mitigation under the Cancun Agreements. Parties then continued to work towards a post-2012 legally-binding agreement with the Cancun Agreements and Bali Road Map as their foundations, and the Convention and KP as their guides. At COP17 in Durban, they reached agreement on a second commitment period on the Kyoto Protocol and on a pathway and deadlines to drawing up and committing to a new, post-2020 mitigation framework under the Convention (see section on Durban Platform for Enhanced Action). All industrialised countries and 48 developing countries also affirmed their pledges up to 2020.[24]

At COP18 in Doha, Qatar, in December 2012, Parties adopted the decision 1/CP.18 (Agreed outcome pursuant to the Bali Action Plan), thereby successfully concluding 5 years of work under the Bali Action Plan, with significant achievements enabling enhanced action on mitigation, adaptation, finance, technology and capacity-building.[25]

Although attempts were made to keep the two tracks separate, each clearly had implications for the other. Developing nations wanted the Kyoto Protocol to continue with more aggressive targets, while other countries, particularly the United States, under the Convention, wanted a new agreement that included all large emitters.

Both working groups were to complete their assignments and report at the 2009 UNFCCC meetings in Copenhagen, Denmark, in the hope that their work would lead to a new agreement. No agreement was reached, however, and the COP and CMP extended the working groups' mandates.

The Copenhagen Accord

The Copenhagen Accord, a political agreement struck by world leaders at the 2009 UN Climate Change Conference in Copenhagen, calls on participating countries to pledge specific actions they will undertake to mitigate greenhouse gas emissions. This represents the first time ever that all of the world's major economies have offered explicit international climate pledges.

The Copenhagen Agreement is a document that delegates at the 15th session of the Conference of Parties (COP 15) to the United Nations Framework convention on Climate Change agreed to "take note of" at the final plenary on 18 December 2009.

The Accord, drafted by, on the one hand, the United States and on the other, in a united position as the BASIC countries (China, India, South Africa, and Brazil), is not legally binding and does not commit countries to agree to a binding successor to the Kyoto Protocol, whose round ended in 2012.[26]

The basic terms of the Copenhagen Accord were brokered directly by President Obama and a handful of key developing country leaders on the final day of the conference, capping two weeks of harsh rhetoric and pitched procedural battles that made the prospect of any agreement highly uncertain. It then took nearly another full day of tense negotiations to arrive at a procedural compromise allowing the leaders' deal to be formalized over the bitter objections of a few governments.[27]

In the end, parties adopted parallel decisions under the UN Framework Convention on Climate Change (UNFCCC) and the Kyoto Protocol that "take note" of the political accord and open the way for governments to individually sign on. In separate decisions, parties extended Ad Hoc Working Groups under both the Convention and the Protocol to continue negotiating toward a fuller agreement in late 2010 in Mexico. The unusual set of outcomes leaves uncertainty, however, about the formal standing of the Copenhagen Accord under the UN climate process and about the nature of any future agreement. The aim of a "legally binding instrument," which appeared part of the deal when President Obama first announced it, was later stripped out.[28]

Key elements of the Copenhagen Accord include: an aspirational goal of limiting global temperature increase to 2°C; a process for countries to enter their specific mitigation pledges by January 31, 2010; broad terms for the reporting and verification of countries' actions; a collective commitment by developed countries for $30 billion in "new and additional" resources in 2010-12 to help developing countries reduce emissions, preserve forests, and adapt to climate change; and a goal of mobilizing $100 billion a year in public and private finance by 2020 to address developing country needs. The accord also calls for the establishment of a Copenhagen Green Climate Fund, a High Level Panel to examine ways of meeting the 2020 finance goal, a new Technology Mechanism, and a mechanism to channel incentives for reduced deforestation.[29]

The Copenhagen Climate Change Conference raised climate change policy to the highest political level. Close to 115 world leaders attended the high-level segment, making it one of the largest gatherings of world leaders ever outside UN headquarters in New York. More than 40,000 people, representing governments, nongovernmental organizations, intergovernmental organizations, faith-based organizations, and media and UN agencies applied for accreditation.

Other dramas in Copenhagen included open squabbling among the typically unified developing country Group of 77, and the struggle between the issue's two lead protagonists– the United States and China. The two sparred before the press and remained deadlocked behind closed doors until nearly the end. Secretary of State Hillary Clinton, who arrived ahead of President Obama, upped the pressure by declaring US support for the goal of $100 billion a year for developing countries, an offer that many African and small-island countries did not want to let slip by. It was only then that Chinese Vice Foreign Minister He Yafei appeared to concede on US demands that its actions be open to some form of international scrutiny.[30]

President Obama closed the deal the next day in a meeting with Chinese Premier Wen Jiabao, Indian Prime Minister Manmohan Singh, Brazilian President Luiz Inacio Lula, and South African President Jacob Zuma, He then announced the tentative agreement to the press and headed home, leaving other leaders to consider the terms, and weary negotiators to devise the final procedural manoeuvres.[31]

COP 15 / CMP 5 was a crucial event in the negotiating process

- Change cooperation, including improvements to the Clean Development Mechanism of the Kyoto Protocol.
- Significant progress was made in narrowing down options and clarifying choices needed to be made on key issues later on in the negotiations.
- It produced the Copenhagen Accord, which expressed clear a political intent to constrain carbon and respond to climate change, in both the short and long term.

The Copenhagen Accord contained several key elements on which there was strong convergence of the views of governments. This included the long-term goal of limiting the maximum global average temperature increase to no more than 2°C above pre-industrial levels, subject to a review in 2015. There was, however, no agreement on how to do this in practical terms. It also included a reference to consider limiting the temperature increase to below 1.5 degree a key demand made by vulnerable developing countries. Other central elements included:

- Developed countries' promises to fund actions to reduce greenhouse gas emissions and to adapt to the inevitable effects of climate change in developing countries. Developed countries promised to provide US$30 billion for the period 2010-12, and to mobilize long-term finance of a further US$100 billion a year by 2020 from a variety of sources.
- Agreement on the measurement, reporting and verification of developing country actions, including a reference to "international consultation and analysis", which had yet to be defined.
- The establishment of four new bodies: a mechanism on REDD-plus, a High-Level Panel under the COP to study implementation of financial provisions, the Copenhagen Green Climate Fund, and a Technology Mechanism.

The work of the two central negotiating groups, the AWG-LCA and the AWG-KP was extended by the COP.

The Accord

The Copenhagen Accord is a political (as opposed to legal) agreement of a novel form. Formal decisions under the UN climate process are typically taken by consensus. As some parties opposed the accord, the decision entering it into the conference's proceedings is not technically an acceptance of its substantive content by the Conference of the Parties (or by the parallel Meeting of the Parties under Kyoto). Rather, the decisions by the two bodies' only "take note" of the attached accord. Individual countries, in all likelihood a strong majority of the Convention's 192 parties, will affix their names to the accord in the coming weeks. The accord declares itself "operational immediately," although many of its provisions will require further elaboration (in some cases explicitly, and in other cases presumably, by the UNFCCC Conference of the Parties). The timeline for doing so is not specified.[32]

In substance, the accord speaks to all of the core elements of the Bali Action Plan: a long-term goal; mitigation; adaptation; finance; technology; forests; and measurement, reporting and verification.[33]

- Endorses the continuation of the Kyoto Protocol.
- Underlines that climate change is one of the greatest challenges of our time and emphasises a "strong political will to urgently combat climate change in accordance with the principle of common but differentiated responsibilities and respective capabilities."
- To prevent dangerous anthropogenic interference with the climate system, recognizes "the scientific view that the increase in global temperature should be below 2°C", in a context of sustainable development, to combat climate change.
- Recognizes "the critical impacts of climate change and the potential impacts of response measures on countries particularly vulnerable to its adverse effects" and stresses "the need to establish a comprehensive adaptation programme including international support."
- Recognizes that "deep cuts in global emissions are required according to science" (IPCC AR4) and agrees cooperation in peaking (stopping from rising) global and national greenhouse

gas emissions "as soon as possible" and that "a low-emission development strategy is indispensable to sustainable development."

- States that "enhanced action and international cooperation on adaptation is urgently required to... reduce vulnerability and build... resilience in developing countries, especially in those that are particularly vulnerable, especially least developed countries (LDCs), small island developing states (SIDS) and Africa" and agrees that "developed countries shall provide adequate, predictable and sustainable financial resources, technology and capacity-building to support the implementation of adaptation action in developing countries."
- About mitigation agrees that developed countries (Annex I Parties) would "commit to economy-wide emissions targets for 2020" to be submitted by 31 January 2010 and agrees that these Parties to the Kyoto Protocol would strengthen their existing targets. Delivery of reductions and finance by developed countries will be measured, reported and verified (MRV) in accordance with COP guidelines.
- Agrees that developing nations (non-Annex I Parties) would "implement mitigation actions" (Nationally Appropriate Mitigation Actions) to slow growth in their carbon emissions, submitting these by 31 January 2010. LDS and SIDS may undertake actions voluntarily and on the basis of (international) support...
- Agrees that developing countries would report those actions once every two years via the UN climate change secretariat, subjected to their domestic MRV. NAMAs seeking international support will be subject to international MRV
- Recognizes "the crucial role of reducing emission from deforestation and forest degradation and the need to enhance removals of greenhouse gas emission by forests", and the need to establish a mechanism (including REDD-plus) to enable the mobilization of financial resources from developed countries to help achieve this.

- Decides to pursue opportunities to use markets to enhance the cost-effectiveness of, and to promote mitigation actions.
- Developing countries, especially these with low-emitting economies should be provided incentives to continue to develop on a low-emission pathway.
- States that "scaled up, new and additional, predictable and adequate funding as well as improved access shall be provided to developing countries... to enable and support enhanced action."
- Agrees that developed countries would raise funds of $30 billion from 2010-12 of new and additional resources.
- Agrees a "goal" for the world to raise $100 billion per year by 2020, from "a wide variety of sources", to help developing countries cut carbon emissions (mitigation). New multilateral funding for adaptation will be delivered, with a governance structure.
- Establishes a Copenhagen Green Climate Fund, as an operating entity of the financial mechanism, "to support projects, programme, policies and other activities in developing countries related to mitigation". To this end, creates a High-Level Panel.
- Establishes a Technology Mechanism "to accelerate technology development and transfer...guided by a country-driven approach."
- Calls for "an assessment of the implementation of this Accord to be completed by 2015. This would include consideration of strengthening the long-term goal", for example to limit temperature rises to 1.5°C.[34,35]

Relation to the UNFCCC and Kyoto—The accord endorses two parallel decisions under the Convention and the Protocol extending the two formal negotiating tracks that existed prior to Copenhagen. Those decisions, however, do not cross-reference the accord. Thus, while some parties will likely look to those negotiating processes to elaborate and fully operationalize the accord, no formal link was established. A set of decisions addressing the core elements of the Bali Action Plan, and a core decision tying them together, were not completed. A draft decision circulated at the time President Obama announced the tentative deal described the intended outcome next year as "a legally binding instrument."[36]

Cancun Climate Agreements

The Cancun Agreements are a set of significant decisions by the international community to address the long-term challenge of climate change collectively and comprehensively over time and to take concrete action now to speed up the global response.

The agreements, reached on December 11 in Cancun, Mexico, at the 2010 United Nations Climate Change Conference represent key steps forward in capturing plans to reduce greenhouse gas emissions and to help developing nations protect themselves from climate impacts and build their own sustainable futures.

Main Objectives of the Agreements

- Establish clear objectives for reducing human-generated greenhouse gas emissions over time to keep the global average temperature rise below two degree.
- Encourage the participation of all countries in reducing these emissions, in accordance with each country's different responsibilities and capabilities to do so.
- Ensure the international transparency of the actions which are taken by countries and ensure that global progress towards the long-term goal is reviewed in a timely way.
- Mobilize the development and transfer of clean technology to boost efforts to address climate change, getting it to the right place at the right time and for the best effect.
- Mobilize and provide scaled-up funds in the short and long term to enable developing countries to take greater and effective action.
- Assist the particularly vulnerable people in the world to adapt to the inevitable impacts of climate change.
- Protect the world's forests, which are a major repository of carbon.

- Build up global capacity, especially in developing countries, to meet the overall challenge.
- Establish effective institutions and systems which will ensure these objectives are implemented successfully.

Cancun Adaptation Framework

Parties adopted the Cancun Adaptation Framework (CAF) as part of the Cancun Agreements at the 2010 Climate Change Conference in Cancun, Mexico (COP 16/ CMP 6). In the Agreements, Parties affirmed that adaptation must be addressed with the same level of priority as mitigation. The CAF is the result of three years of negotiations on adaptation under the AWG-LCA that had followed the adoption of the Bali Action Plan at the 2007 Climate Change Conference in Bali, Indonesia (COP 13/ CMP 3).[37]

The Cancun Adaptation Framework includes the following five clusters:

1. *Implementation*
   - All Parties to plan, prioritize and implement adaptation actions and to use existing channels to provide information on support provided and received for adaptation actions and on activities undertaken;
   - A process to enable LDC Parties-building upon their experience with the NAPAs—to formulate and implement national adaptation plans and an invitation to other developing country Parties to employ the modalities formulated to support those plans;
   - A work programme to consider approaches to address loss and damage associated with climate change impacts in developing countries that are particularly vulnerable to the adverse effects of climate change.

2. *Support*
   - Developed country Parties to provide developing country Parties, taking into account the needs of those that are particularly vulnerable, with long-term, scaled-up, predictable, new and additional finance, technology, and capacity-building (paras 95-137) to implement

adaptation actions, plans, programmes and projects at local, national, sub-regional and regional levels, including activities under the Cancun Adaptation Framework.

3. *Institutions*
    - **At the global level**: establishment of an Adaptation Committee to promote the implementation of enhanced action on adaptation in a coherent manner under the Convention;
    - **At the regional level**: strengthening and, where necessary, establishing regional centres and networks, in particular in developing countries;
    - **At the national level**: strengthening and, where necessary, establishing and/or designation of national-level institutional arrangements.

4. *Principles*
    - Be undertaken in accordance with the Convention;
    - Follow a country-driven, gender-sensitive, participatory and fully transparent approach, taking into consideration vulnerable groups, communities and ecosystems;
    - Be based on and guided by the best available science and, as appropriate, traditional and indigenous knowledge;
    - Be undertaken with a view to integrating adaptation into relevant social, economic and environmental policies and actions.

5. *Stakeholder engagement*
    - Relevant multilateral, international, regional and national organizations, the public and private sectors, civil society and other relevant stakeholders are invited to undertake and support enhanced action on adaptation at all levels.[38]

The Paris Agreement

On 5 October 2016, the threshold for entry into force of the Paris Agreement was achieved. The Paris Agreement entered into force on 4 November 2016. The first session of the Conference of the Parties serving as the Meeting of the Parties to the Paris Agreement (CMA 1) took place in Marrakech, Morocco from 15-18 November 2016.

At COP 21 in Paris, on 12 December 2015, Parties to the UNFCCC reached a landmark agreement to combat climate change and to accelerate and intensify the actions and investments needed for a sustainable low carbon future. The Paris Agreement builds upon the Convention and–for the first time–brings all nations into a common cause to undertake ambitious efforts to combat climate change and adapt to its effects, with enhanced support to assist developing countries to do so. As such, it charts a new course in the global climate effort.[39]

The Paris Agreement's central aim is to strengthen the global response to the threat of climate change by keeping a global temperature rise this century well below 2°C above pre-industrial levels and to pursue efforts to limit the temperature increase even further to 1.5°C. Additionally, the agreement aims to increase the ability of countries to deal with the impacts of climate change, and at making finance flows consistent with a low GHG emissions and climate-resilient pathway. To reach these ambitious goals, appropriate mobilization and provision of financial resources, a new technology framework and enhanced capacity-building is to be put in place, thus supporting action by developing countries and the most vulnerable countries, in line with their own national objectives. The Agreement also provides for an enhanced transparency framework for action and support.[40]

The Paris Agreement opened for signature on 22 April 2016 – Earth Day – at UN Headquarters in New York. It entered into force on 4 November 2016, 30 days after the so-called "double threshold" (ratification by 55 countries that account for at least 55 per cent of global emissions) had been met. Since then, more countries have ratified and continue to ratify the Agreement, reaching a total of 125 Parties in early 2017. The current number of ratifications can be found here.[41]

In order to make the Paris Agreement fully operational, a work programme was launched in Paris to develop modalities, procedures and guidelines on a broad array of issues. Since 2016, Parties work together in the subsidiary bodies (APA, SBSTA and SBI) and various constituted bodies. The Conference of the Parties serving as the meeting of the Parties to the Paris Agreement (CMA) met for the first time in conjunction with COP 22 in Marrakesh (in November 2016) and adopted its first two decisions. The work programme is expected to be completed by 2018.[42]

Global warming may indeed be the sharpest example of how policy in Washington will change under a Trump administration. President Obama has said his efforts to establish the United States as the global leader in climate policy are his proudest legacy.

Trump has already vowed to 'cancel' last year's Paris climate agreement, which commits more than 190 countries to reduce their emissions of planet-warming carbon dioxide pollution, and to dismantle the Clean Power Plan. Though it is not very easy to change Paris climate change agreement; the impact of global warming might change Trump's mind.

"Trump's election, in a moment when climate action has finally been mainstreamed into the economy, gives Republicans a chance to reconcile with real life," André Ferretti, a coordinator of the Brazilian Climate Observatory, said. "In real life, we keep breaking temperature records, and extreme events impact Democrats and Republicans alike."[43]

India also committed during the Paris talks to increasing forest cover. Prakash Javadekar, the minister of environment, forest and climate change, has said publicly in recent weeks that the government planned to commit 400 billion Indian rupees, or about $6 billion, toward reforestation programs.

Criticism

A pair of studies in Nature have said that, as of 2017, none of the major industrialized nations were implementing the policies they had envisioned and have not met their pledged emission reduction targets, and even if they had, the sum of all member pledges (as of 2016) would not keep global temperature rise "well below 2°C". According to the UNEP the emission cut targets in November 2016 will result in temperature rise by 3°C above pre-industrial levels, far above the 2°C of the Paris climate agreement.

In addition, an MIT News article written on 22 April 2016 discussed recent MIT studies on the true impact that the Paris Agreement had on global temperature increase. Using their Integrated Global System Modelling (IGSM) to predict temperature increase results in 2100, they used a wide range of scenarios that included no effort towards climate change past 2030, and full extension of the Paris Agreement past 2030. They concluded that the Paris Agreement would cause temperature decrease by about 0.6 to 1.1°C compared to a no-effort-scenario, with only a 0.1°C change in 2050 for all scenarios. They concluded that, although beneficial, there was strong evidence that the goal provided by the Paris Agreement could not be met in the future; under all scenarios, warming would be at least 3.0°C by 2100.

How well each individual country is on track to achieving its Paris agreement commitments can be continuously followed on-line (through the Climate Action Tracker and the Climate Clock).

2018 published study points at a threshold at which temperatures could rise to 4 or 5 degree compared to the pre-industrial levels, through self-reinforcing feedbacks in the climate system, suggesting this threshold is below the 2-degree temperature target, agreed upon by the Paris climate deal. The author of University of Copenhagen Katherine Richardson's study, stresses, "We note that the Earth has never in its history had a quasi-stable state that is around 2°C warmer than the pre-industrial and suggest that there is substantial risk that the system, itself, will 'want' to continue warming because of all of these other processes – even if we stop emissions. This implies not only reducing emissions but much more."

At the same time, another 2018 published study notes that even at a 1.5°C level of warming, important increases in the occurrence of high river flows would be expected in India, South and Southeast Asia. Yet, the same study points out that under a 2.0°C of warming various areas in South America, central Africa, western Europe, and the Mississippi area in the United States would see more high flows; thus increasing flood risks.

Institutional asset owners associations and think-tanks have also observed that the stated objectives of the Paris Agreement are implicitly "predicated upon an assumption – that member states of the United Nations, including high polluters such as China, the US, India, Russia, Japan, Germany, South Korea, Iran, Saudi Arabia, Canada, Indonesia and Mexico, which generate more than half the world's greenhouse gas emissions, will somehow drive down their carbon pollution voluntarily and assiduously without any binding enforcement mechanism to measure and control CO_2 emissions at any level from factory to state, and without any specific penalty gradation or fiscal pressure (for example a carbon tax) to discourage bad behaviour."

According to United Nations Environment Programme (UNEP) if we rely only on the current climate commitments of the Paris Agreement, temperatures will likely rise to 3.2°C this century. To limit the global temperature rise to 1.5°C, emissions must be below 25 gigatons (Gt) by 2030. With current Nov 2019 commitments, emissions will be 56 Gt CO_2 by 2030, twice the environmental target. To limit global temperature rise to 1.5°C, the global annual emission reduction needed is 7.6 per cent emissions reduction every year between 2020 and 2030. The top four emitters (China, USA, EU28 and India) contribute to over 55 per cent of the total emissions over the last decade, excluding emissions from land-use change such as deforestation. China's emissions grew 1.6 per cent in 2018 to reach a high of 13.7 Gt of CO_2 equivalent. The US emits 13 per cent of global emissions and emissions rose 2.5 per cent in 2018. The EU emits 8.5 per cent of global emissions has declined 1 per cent per year across the last decade. Emissions declined 1.3 per cent in 2018. India's 7 per cent of global emissions grew 5.5 per cent in 2018. An emission per capita is one of the lowest within the G20.

NOTES

1. Williams, Tim, Climate Change Negotiations: The United Nations Framework Convention on Climate Change in Context, Publication Library of Parliament Ottawa, Canada, 2015, p. 1
2. Ibid, p. 2
3. Ibid.
4. Ibid.
5. Ibid.
6. Ibid.
7. Ibid, p. 5

8. Ibid.
9. Ibid, p. 9.
10. Ibid.
11. The Kyoto Protocol: History, Accessed through https://www.mtholyoke.edu/~danov20d/site/history.htm on 25 December 2016.
12. Ibid.
13. Ibid.
14. United Nations Framework Convention on Climate Change, Accessed through http://unfccc.int/kyoto_protocol/background/items/2879.phpp.1, on 30 December2016.
15. Ibid.
16. United Nations Framework Convention on Climate Change, Accessed through http://unfccc.int/kyoto_protocol/background/items/2879.phpp. 2, on 30 December 2016.
17. Ibid.
18. Ibid.
19. Kyoto Protocol Fast Facts, CNN Library Updated 1712 GMT (0112 HKT) March 30, 2016 Accessed through http://edition.cnn.com/2013/07/26/world/kyoto-protocol-fast-facts/ on 1 January 2017.
20. United Nations Framework Convention on Climate Change, Accessed through http://unfccc.int/key_steps/bali_road_map/items/6072.php on 3 January2017.
21. Ibid.
22. Ibid.
23. Ibid.
24. United Nations Framework Convention on Climate Change, Accessed through http://unfccc.int/key_steps/bali_road_map/items/6072.php on 5 January2017.
25. Ibid.
26. Copenhagen Accord: Accessed through http://unfccc.int/resource/docs/2009/cop15/eng/107.pdfon 6 January 2017.
27. Centre For Climate and Energy solutions, Working Together for the Environment and the Economy, Accessed through https://www.c2es.org/international/negotiations/cop-15/summary Fifteenth Session of the Conference of the Parties to the United Nations Framework Convention on Climate Change and Fifth Session of the Meeting of the Parties to the Kyoto Protocol, December 7-18, 2009Copenhagen, Denmark on 7 January2017.
28. Ibid.
29. Ibid.
30. Ibid.
31. Ibid.
32. Centre For Climate and Energy solutions, Working Together for the Environment and the Economy, Accessed through https://www.c2es.org/international/negotiations/cop-15/summary Fifteenth Session of the Conference of the Parties to the United Nations Framework Convention on Climate Change and Fifth Session of the Meeting of the Parties to the Kyoto Protocol, December 7-18, 2009Copenhagen, Denmark on 7 January2017.
33. Ibid.

34. Copenhagen Accord: Accessed through http://unfccc.int/resource/docs/2009/cop15/eng/107.pdfon 6 January 2017.
35. Factbox, What was agreed and left unfinished in UN climate deal Accessed through http://in.reuters.com/article/idINIndia-44872920091220?sp=true on 6January 2017.
36. Centre for Climate and Energy solutions, Working Together for the Environment and the Economy. Accessed through https://www.c2es.org/international/negotiations/cop-15/summary Fifteenth Session of the Conference of the Parties to the United Nations Framework Convention on Climate Change and Fifth Session of the Meeting of the Parties to the Kyoto Protocol, December 7-18, 2009Copenhagen, Denmark on 7 January2017.
37. United Nations Framework Convention on Climate Change, Accessed through http://unfccc.int/adaptation/items/5852.phn on 6 January 2017.
38. United Nations Framework Convention on Climate Change, Accessed through http://unfccc.int/adaptation/items/5852.phn on 6 January 2017.
39. United Nations Framework Convention on Climate Change, Summary of the Paris Agreement, Accessed through http://bigpicture.unfccc.int/#content-the-paris-agreement on 7 January 2017.
40. Ibid.
41. Ibid.
42. Ibid.
43. At U.N. Meeting, Diplomats Worry Trump Could Cripple Climate Pact. By CORAL DAVENPORT 15 NOV2016. Accessed through https://www.nytimes.com/2016/11/16/world/united-nations-climate-change-trump.html on 8 January2017.

List of Figures, Chart and Tables

Bibliography

Books and Journal Articles

1. Adger, Neil W.; Paavola, Jouni; Huq, Saleemul; M.J., Mace, *Fairness in Adaptation to Climate Change,* MIT Press, Cambridge, 2006.
2. Armstrong, Sinnott Walter; Howarth, B. Richards, *Advances in the Economics of Environmental Resources, Vol. 5,* Jai Press, 2005.
3. Ausubel, J.H., *Does Climate Still Matter?, Nature* 350, pp.649-652, 1991.
4. Air Commodore (Retd.) Jasjit Singh, "Future Directions of India's Defence Policy", *Asian Defence Journal,* pp. 28-29, November 1995.
5. Baettig, M.B.; Brander, S. and Imboden, D.M., "Measuring Countries' Cooperationwithin the International Climate Change Regime", *Environmental Science & Policy* 11, pp. 478-489, 2008
6. Baer P., T. Athanasiou, S. Kartha, and E. Kemp-Benedict, *Greenhouse Development Rights: A proposal for a fair global climate treaty, Ethics, Place & amp., Environment* 12, pp. 267–281, 2009.
7. Barnett J., "Security and Climate Change," *Global Environmental Change,* 13, 7-17, 2003.
8. Barnett J., *Security and Climate Change,* Tyndall Centre for Climate Change Research Tyndall, Centre for Climate Change Research Working Paper 7, 2001
9. Barrieu P., and B. Sinclair-Desgagné, *On Precautionary Policies, Management Science* 52 pp. 1145 –1154, 2006.
10. Baldwin D.A., *The Concept of Security Review of International Studies,* Vol. 23, Issue 1, pp. 5-26, January 1997.
11. Bodansky, D., "Deconstructing the Precautionary Principle," in *Bringing New Law to Ocean Waters,* (Eds.) D.D. Caron, H.N. Scheiber, Brill, Leiden, Netherlands, 2004
12. Bolin, Bert, *A History of the Science and Politics of Climate Change: The Role of the Intergovernmental Panel on Climate Change.* Cambridge University Press, 15 November 2007.

13. Bodansky, D., *Legal Options for U.S. Acceptance of a New Climate Change Agreement,* Sandra Day O'Connor College of Law, Arizona State University, p. v, May 2015.
14. Bosetti, V., Carraro, C., Sgobbi, A., Tavoni M., *Modeling economic impacts of alternative international climate policy architectures: A quantitative and comparative assessment of architectures for agreement. In: Post-Kyoto International Climate Policy: Implementing Architectures for Agreement,* (Eds.) J.E. Aldy, R.N. Stavins, Cambridge University Press, Cambridge, UK, pp. 715-752, 2010.
15. Brodie, J., "From Social Security to Public Safety: Security Discourses and Canadian Citizenship," *University of Toronto Quarterly,* Vol. 78, pp. 687-708 (Article), 2009.
16. International Climate Change Regime, Environmental Science; Policy 11, pp. 478-48.
17. Boykoff, T. Maxwell, *The Politics of Climate Change A Survey,* Taylor & Francis Ltd, 2010.
18. Bailey, I. Compston, H., *Feeling the Heat: The Politics of Climate Policy in Rapidly Industrializing,* Palgrave Macmillan Basingstoke, UK, 2012.
19. Bryner, G., *Failure and Opportunity: Environmental Groups in US Climate Change Policy,* Online Publishing Taylor & Francis, 2008.
20. Boyce, T., Lewis, Justin, *Climate Change and the Media,* Peter Lang Publishing INC, New York, 2009.
21. Brunner, R., Lynch, A., *Adaptive Governance and Climate Change,* American Meteorological Society Massachusetts, 2013.
22. Bomberg, E., Schlosberg, D., *Environmentalism in the United States: Changing Conceptions of Activism,* Routledge Oxon and New York, 2013.
23. Braasch, G., McCabe, B., *Earth Under Fire: How Global Warming is Changing the World,* University of California Press, California, 2007.
24. Brown, A. Marilyn, Sovaco, K. Benjamin, *Climate Change and Global Energy Security: Technology and Policy Options,* The MIT Press, England, 2011.
25. Cao, J., *Reconciling Human Development and Climate Protection: A multistage hybrid climate policy architecture. In: Post-Kyoto International Climate Policy: Implementing Architectures for Agreement,* (Eds.) J.E. Aldy, R.N. Stavins, Cambridge University Press, Cambridge, UK, pp.563–598, 2010a.
26. Chatterjee, A., *Global Warming and Climate Change,* Global Publications, Delhi, 2010.
27. Cohen R., Rai, M Shirin, *Global Social Movements,* The Athlone Press, 2000.
28. Crow, D., Boykoff, M., *Culture, Politics and Climate Change: How Information Shapes our Common Future,* Routledge, Oxon & New York, 2014.

29. Compston, H., Bailey, I., *Climate Clever: How Governments Can Tackle Climate Change (and Still Win Elections)*, Routledge, Oxon and New York, 2013.
30. Campbell, M. Kurt., *Climatic Cataclysm: The Foreign Policy and National Security Implications of Climate Change*, Brookings Institute Press, Washington DC, 2009.
31. Dessler, A., A. Parson, Edward, *The Science and Politics of Global Climate Change: A Guide to the Debate*, 2nd Edition, Cambridge University Press, 2009
32. Dalby, S., *Ecopolitical Discourse: Environmental Security and Political Geography*, Progress in Human Geography, pp. 503-522, 1992.
33. Dessler, Andrew Parson, A, Edward, *The Science and Politics of Global Climate Change: A Guide to the Debate*, Cambridge University Press, 2010.
34. Doyle, J., *Mediating Climate Change*. Ashgate Publishing Limited, England, Ashgate Publishing University, USA, 2011.
35. Downie, Leonard David; Brash, Kate; Vaughan, Catherine, *Climate Change: A Reference Handbook*, ABC-CLIO Inc, 2009.
36. Downie, C., *The Politics of Climate Change Negotiations: Strategies and Variables in Prolonged International Negotiation (New Horizon of International Politics)*, Edwards Elgar Publishing Limited, UK, 2014.
37. Dubash, N., *Handbook of Climate Change and India: Development, Politics and Governance*. Earthscan Oxon, USA & Canada, 2012.
38. Davoudi, S.; crawford, J.; Mehmood, A., *Planning for Climate Change: Strategy for Mitigation and Adaptation for Spatial Planners*, Earthscan, UK & USA, 2009.
39. Dingell, J., *Perspectives on Climate Change*, US Government Printing Office, 2008.
40. Downie, Leonard David; Brash, Kate; Vaughan, Catherine, *Climate Change: A Reference Handbook*, ABC-CLIO, England, 2009.
41. Dinar, A., *Measuring the Impact of Climate Change on Indian Agriculture*, The world Bank, Washington DC, 1998.
42. Dankelman, I., *Gender and Climate Change: An Introduction*, Earthscan, London, 2010.
43. Emanuel, K., *What We Know About Climate Change*, MIT Press, Cambridge, England 2012.
44. Environment Protection Agency EPA, *Greenhouse Gas and Climate Change*, Sweden, 1983.
45. Ereaut, G.; Segnit, N., *Warm Words: How are we telling the Climate Story and can we tell it better?*, Technical Report, Institute for Public Policy Research, London, p. 32, 2006.
46. Feigenbaum, Evan A., "India's Rise, America's Interest: The Fate of the U.S.-Indian Partnership, Council on Foreign Relations," *Foreign Affairs*, Vol. 89, No. 2, pp. 76-91, March/April 2010.

47. Fickling, M.; Schott, J. Jeffrey, *NAFTA and Climate Change*, Versa Press, USA, 2011.
48. Finan, F. William, *Analysis of the Effects of U.S. National Security Controls on U.S: Headquartered Industrial Firms*, Vol. 1 Washington DC, August 25 1986.
49. Giddens, A., *The Politics of Climate Change*, Polity Press Reprinted, 2009.
50. Giddens, A., *Politics of Climate Change: A European Perspective*, Routledge, 1996.
51. Giddens, A., *The Politics of Climate Change*, Wiley Online Library Cambridge, UK, 2009.
52. Gaan, N., *Climate Change and International Politics*, Kalpaz Publications, Delhi, 2008.
53. Goodal, C., *Ten Technologies to Fix Energy and Climate*, Green Profile Books Ltd, London, 2010.
54. Goldstone, J., *Demography, Environment, and Security*, (Eds.) Diehl, P., Gleditsch, N., *Environmental Conflict*, Westview Press, Boulder, pp. 84-108, 2001.
55. Gilpin, Robert, *War and Change in World Politics*, University of Cambridge, 1983.
56. Ganguly, S., Scobell, A., Shoup, B., *US-Indian Strategic Cooperation into the 21st Century: More than Words*, Routledge, Oxon & New York, 2006.
57. Geri, R. Laurance, McNab, E. David, *Energy Policy in the U.S.: Politics, Challenges, and Prospects for Change*, CRC Press, Boca Raton, 2016.
58. Green, J.M., Freeman, W.C., Searight, E. Amy, *Green Dragons: The Politics of Climate Change in Asia*, Centre for Strategic and International Studies Library of Congress Cataloging in Publication Data, Washington DC, November, 2010
59. Gerrard, Michael, *Global Climate Change and U.S. Law*, Section of Environment Energy and Resources Book Publication Committee, 2006-2007.
60. Hergerl, G.C., Zwiers, F.W., Braconnot, P., Gillet, N.P., Luo, Y., Marengo Orsini, J.A., Nicholls, N., Penner, J.E. and Stott, P.A., *Understanding and Attributing Climate Change*. In: *Climate Change: The Physical Science Basis. Contribution of Working Group I to the Fourth Assessment Report of the Intergovernmental Panel on Climate Change*, 2007.
61. Houghton, J., *Global Warming*, Cambridge University Press, Houghton, 1994 1997 2004.
62. Hammitt J., *Stratospheric Ozone Depletion and Global Climate Change*. In: *The Reality of Precaution: Comparing Risk Regulation in the United States and Europe*. (Eds) J. Wiener, M. Rogers, J. Hammitt, P. Sand, RFF Press, Washington DC, 2010.
63. Hoegh-Guldberg, O., Hoegh-Guldberg, H., Stout, D., Cesar,

H.,Timmerman, A., *Pacific in Peril: Biological, Economic and Social Impacts of Climate Change on Pacific Coral Reefs. Greenpeace*, Amsterdam, 2000.

64. Hardy T. John, *Climate Change Causes, Effects and Solutions*. John Wily & Sons Ltd, The Atrium Southern Gate Chichester, England, 2003.
65. Howe, P. Joshua, *Behind the Curve: Science and the Politics of Global Warming*, University of Washington Press, 2014.
66. Hahn,William Robert: *The Economics and Politics of Climate Change*, AEI Press, 1998.
67. Helm, D., Hepburn, C., *The Economics and Politics of Climate Change*, OUP Oxford, 2009.
68. Harris, G.P., *Climate Change and Foreign Policy: Case Studies from East to West*, Routledge, London and New York, 2009.
69. Harris, G.P., *The Politics of Climate Change: Environmental Dynamics in International Affairs*, Routledge Oxon and New York, 2012.
70. Hollo, E., Kulovesi, K., Mehling, M., *Climate Change and New Law*, Springer Dordrecht Heidelberg, New York, London, 2013.
71. Harrison, K., Sundstrom, L., *Global Commons, Domestic Decisions: The Comparative Politics of Climate Change*, MIT Press, Cambridge England, 2010.
72. Heidelberg, Berlin, *World in Transition: Climate Change as a Security Risk*, Springer, WBGU, 2007.
73. Hale, *S., The New Politics of Climate Change: Why we are failing and how we will succeed Book*, Environmental Politics, Taylor & Francis, 2010.
74. Hulme, M., *Why we Disagree about Climate Change: Understanding Controversy, Inaction and Opportunity*, Cambridge University Press, 2009.
75. Hardy, T. J., *Climate Change: Causes, Effects, and Solutions*, John Weily & Sons England, 2003.
76. Hastrup, K., Skrydstrup M., *The Social Life of Climate Change Models: Anticipating Nature*, Routledge, New York & UK, 2013.
77. Ikeme, J., "Equity, Environmental Justice and Sustainability: Incomplete Approaches in Climate Change Politics," *Global Environmental Change*, 13, pp. 195-206, 2003.
78. IPCC, 2007: *Climate Change: Mitigation of Climate Change: Contribution of Working Group III to the Fourth Assessment Report of the Intergovernmental Panel on Climate Change*, Cambridge University Press, Cambridge, UK & New York, 2007.
79. IPCC, 2001a: *Technical Summary: Climate Change impacts, adaptation, and vulnerability, in Climate Change 2001: Impacts, Adaptation, and Vulnerability. Contribution of Working Group II to the Third Assessment Report of the Intergovernmental Panel on Climate Change*, Cambridge University Press, Cambridge, 2001
80. IPCC, 2013: *Summary for Policymakers*. In: *Climate Change 2013: The*

114. Neufeldt, H., *Making Climate Change Work for Us: European Perspectives on Adaptation and Mitigation Strategies,* Cambridge University Press, 2009.

115. Okereke, C., "Climate Justice and the International Regime," *Wiley Interdisciplinary, Reviews: Climate Change,* 1, pp. 462–474, 2010.

116. O'riordan, T., *Politics of Climate Change: A European Perspective,* Routledge London & Routledge New York British Library, 1996.

117. Parry, M., "Viewpoint: Climate Change—Where Should our Research Priorities be?," *Global Environmental Change,* 11, pp. 257-260, 2001.

118. Paterson, M., *Global Warming and Global Politics,* Routledge, London UK, 1996

119. Posner, E., Sunstein, C., *Justice and Climate Change: The unpersuasive case for per capita allocations of emissions rights. In: Post-Kyoto International Climate Policy:Implementing Architectures for Agreement.* In edited by J.E. Aldy, R.N. Stavins, Cambridge University Press, Cambridge, UK, pp. 343–371, 2010.

120. Pittock, Barrie A., *Climate Change: The Science, Impacts and Solutions.* Earthscan, 2009.

121. Planning Commission, Government of India: *An Interim Report on Low Carbon 1, Strategies for Inclusive Growth,* 2011.

122. Publisher, Congress, *United States Congressional Serial Set United States,* US Government Printing Office, 2003.

123. Pataki, E. George, Vilsack, J. Thomas, Levi, A. Michael, *Confronting Climate Change: A Strategy for U.S. Foreign Policy,* Council on Foreign Relations, USA, 2008.

124. Putra, Azha Nur, Han, Eulalia, *Governments Responses to Climate Change: Selected Examples from Asia Pacific,* Springer Singapore Heidelberg, New York, London, 2013.

125. Peterson, L. David, Vose, M. James, *Climate Change and United States Forests,* Springer Singapore Heidelberg, New York, London, 2014.

126. Perkovich, George, *India's Nuclear Bomb: The Impact on Global Proliferation,* University of California Press, UK, 2001.

127. Pascual, C., Elkind, J., *Energy Security: Economics, Politics, Strategies, and Implications,* Brookings Institution Press, Washington DC, 2010.

128. Rabe, G. Barry, *Greenhouse Governance: Addressing Climate Change in America,* Brookings Institution Place, Washington DC, 2010.

129. *Regional Study on Greenhouse Effect and its Impact on the Eegion; Kathmandu, SAARC, Secretariat,* December 1992.

130. Richardson, K., Steffen, W., Liverman, D., *Climate Change: Global Risks, Challenges and Decisions,* Cambridge University Press, USA, 2011.

131. Rabe, B.G., *Statehouse and Greenhouse: The Emerging Politics of American Climate Change Policy,* Routledge, London, UK, 2004.

132. Rao M. Vijayalakshmi, *Challenges of Peace and Security in the 21st Century Third Concept,* Vol. 131, No. 11, pp.7-9. January 1998.
133. Rapkin, D., Avery, W., "World Markets and Political Instability within less Developed Countries," *Co-operation & Conflict,* 21(2), pp. 99-117, 1986.
134. Rayner, S., "How to eat an elephant: A bottom-up approach to climate policy," *Climate Policy,* 10, pp. 615–621, 2010.
135. Rowlands, I., *The Politics of Global Atmospheric Change,* Manchester University Press, 1995.
136. Rong F., "Understanding Developing Country Stances on Post-2012 Climate Change Negotiations: Comparative Analysis of Brazil, China, India, Mexico and South Africa," *Energy Policy,* 38, pp. 4582-4591, 2010.
137. Read, P., *Responding to Global Warming: The technology, economics and politics of sustainable energy,* Zed Books Ltd, London & New Jersey, 1994.
138. Roaf, S., Crichton, D., Nicol, F., *Adapting Buildings and Cities for Climate Change: A 21st Century Survival Guide,* Architectural Press, UK & USA, 2009.
139. Rayfuse, G. Rosemary, Scott, V. Shirley, *International Law in the Era of Climate Change,* Edward Elgar Publishing Limited, New South Wales Australia, 2012.
140. Rahm, D., *Climate Change Policy in the United States: The Science, the Politics and the Prospects for Change,* McFarland & Company Inc Publishers, North Carolina, 2009.
141. Rothe, D., *Securitizing Global Warming: A Climate of Complexity,* Routledge, Oxon & New York, 2016.
142. Rajamani, Lavanya, "India and Climate Change: What India Wants, Needs, and Needs to Do," *India Review,* Vol. 8, No. 3, pp. 340-374, July-September, 2009.
143. Sathaye J., Shukla P.R. and Ravindranath N.H., "Climate Change, Sustainable Development and India: Global and National Concerns," *Current Science,* Vol.90, No. 3, pp. 314-325, 2006.
144. Sturges, W.T., *Extension of the Heat Trapping Gases and Aerosols: A Potent Greenhouse Gas Identified in the Atmosphere,* SF5CF3-Science, 289, pp. 611-613, 2000.
145. Singh Jasjit, *India's Strategic and Security Perspective,* Vol. XIII, No. 5, pp. 475-500, August 1990.
146. Sen A., *The Idea of Justice,* Belknap Press of Harvard University Press, Cambridge, MA, 2009.
147. Somanathan, E., *What do we expect from an international climate agreement? A perspective from a low-income country. In: Post-Kyoto International Climate Policy: Implementing Architectures for Agreement.* In edited by J.E. Aldy, R.N. Stavins, Cambridge University Press, Cambridge, UK pp. 599–61, 2010.

148. Stern N., *The Economics of Climate Change: The Stern Review*, Cambridge University Press, Cambridge, UK, 2007.

149. Schneider, B., Nocke, T., *Image Politics of Climate Change: Visualizations, Imaginations,Ddocumentations*, transcript Verlag Bielefeld, 2014.

150. Selin, H., VanDeveer, D. Stacy, *Changing Climate in North American Politics: Institutions, Policymaking and Multilevel Governance*, The MIT Press, Cambridge England, 2009.

151. Stripple, J., Bulkeley, H., *Governing the Climate: New Approaches to Rationality, Power and Politics*, Cambridge University Press, 2013.

152. Sovacool, K. Benjamin, *The Routledge Handbook of Energy Security*, British Library, 2010.

153. Selin, H., Van Deveer, D., *Political Science and Prediction: What's next for US Climate Change Policy?*, Blackwell Publishing Inc, 2007.

154. Shareman, D., Smith, Wayne J., *The Climate Change Challenge and the Failure of Democracy*, Praeger Publisher Westport, 2007.

155. Scott, D., Hall, C.M., Stefan, G., *Tourism and Climate Change: Impacts, Adaptation and Mitigation*, Routledge Oxon, 2012.

156. Stern, N., Bowen, A., Whalley, J., *The Global Development of Policy Regimes to Combat Climate Change*, World Scientific Publishing Ltd Singapore, 2014.

157. Stern, Ncholas, Stern, Herbert Nicholas, *Why Are We Waiting?: The Logic, Urgency, and Promise of Tackling Climate Change*, MIT Press Cambridge England, 2015.

158. Selin, H., VanDeveer, D. Stacy, *Changing Climates in North American Politics: Institutions, Policymaking and Multilevel Governance*, MIT Press Cambridge England, 2009.

159. Schneide, H. Stephen, *Climate Change Science and Policy*, Island Press 14, p. 554, December 2009

160. Stern, N., Bowen, A., Whalley, J., *The Global Development of Policy Regimes to Combat Climate Change*, World Scientific Publishing Co Pvt Ltd, 2014.

161. Scheffran, J., Brzoska, M., Brauch, G. Hans, Link, Michael P., Schilling, Janpeter, *Climate Change, Human Security and Violent Conflict: Challenges for Societal Stability*, Springer Singapore Heidelberg, New York London, 2012.

162. Thompson, L., "Climate Change: The Evidence and Our Options," *The Behavior Analyst*, No. 2, 33, pp. 153-170, 2010.

163. Tol, R.S.J., Downing, T.E., Kuik, O.J., Smith. J.B., "Distributional Aspects of Climate Change Impacts," *Global Environmental Change*, 14, pp. 259-272, 2004

164. Toyota, T., Fujikura, R., *Climate Change Mitigation and Development Cooperation*, Routledge, London & New York 2012.

165. Tubi A., Fischhendler I., Feitelson E., *The Effect of Vulnerability on Climate Change Mitigation Policies. Global Environmental Change*, 22, pp. 472-482, 2012.
166. Urueña R., "Risk and Randomness in International Legal Argumentation, Leiden," *Journal of International Law*, 21, pp. 787–822, 2008.
167. Victor, G. David, *Climate Change: Debating America's Policy Options*, Council on Foreign Relations, USA, 2004.
168. Vanderheiden, S., Barry, John, *Political Theory and Global Climate Change*, MIT Press, 2008.
169. Waltz, K., *Man, the State and War*, New York: Columbia University Press, 1959.
170. Webersik, C., *Climate Change and Security: A Gathering Storm of Global Challenges*, Praeger ABC-CLIO, California, 2010.
171. Wisener. J., Law. D., Bodansky, J. Brunnée, E. Hey, *Precaution. In: The Oxford Handbook of International Environmental*, Oxford University Press, New York, pp.597-612, 2007.
172. Williams, T., *Climate Change Negotiations: The United Nations Framework Convention on Climate Change in Context*, Publication Library of Parliament Ottawa, Canada, p. 1, 2015.
173. World Bank, *World Development Report 2010: Development and Climate Change*, The International Bank for Reconstruction and Development, Washington, DC, 2010
174. World Bank, *Cities, Sea, and Storms: Managing Change in Pacific Island Economies*, Washington, 2000a.
175. Young, Oran R., *International Cooperation: Building Regimes for Natural Resources and the Environment*, Ithaca, NY: Cornell University Press, 1989.
176. Zak, M., *American Global Challenges: The Obama Era*, Palgrave Macmillan, 2011.
177. Zhongping, F., Peral, L., *Global Security in a Multipolar World*, Institute for Security Studies European Union, 2009.

Websites and other Sources

1. Air Enforcement: Stationary Sources. Accessed through https://www.epa.gov/enforcement/air-enforcement
2. At U.N. Meeting, Diplomats Worry Trump Could Cripple Climate Pact By CORAL DAVENPORT NOV. 15, 2016, Accessed through https://www.nytimes.com/2016/11/16/world/united-nations-climate-change-trump.html Agarwal, A.; Narain, S.: Global Warming in an Unequal World. A case of environmental colonialism. Accessed through http://cseindia.org/challenge_balance/readings/GlobalWarming%20 Book.pdf

3. An 'Indian exception' and India-US partnership on climate change: Accessed through https://www.brookings.edu/blog/planetpolicy/2015/01/12/an-india-exception-and-india-u-s-partnership-on-climate-change/
4. A historic commitment to protecting the environment and addressing the impacts of climate change. Accessed through https://obamawhitehouse.archives.gov/the-record/climate
5. By the Numbers: The US's warmest year yet. Accessed through http://www.popsci.com/science/article/2013-01/numbers-uss-warmest-year-yet-infographic
6. Burns R. Nicholas, America's Strategic Opportunity with India: The New U.S.-India, Partnership Council on Foreign Relations, Foreign Affairs, Vol. 86, No. 6 Nov.-Dec., 2007, pp. 131-146, under 'Missed Opportunities'
7. Center For Climate and Energy solutions, Working Together for the Environment and the Economy, Accessed through https://www.c2es.org/international/negotiations/cop-15/summary Fifteenth Session of the Conference of the Parties to the United Nations Framework Convention on Climate Change and Fifth Session of the Meeting of the Parties to the Kyoto Protocol, December 7-18, 2009 Copenhagen, Denmark
8. Copenhagen Accord: Accessed through http://unfccc.int/resource/docs/2009/cop15/eng/l07.pdf
9. Clean Water Act Compliance Monitoring: Accessed through https://www.epa.gov/compliance/clean-water-act-cwa-compliance-monitoring
10. Climate change a threat to security, food and humankind—IPCC report. Accessed through https://www.theguardian.com/environment/2014/mar/31/climate-change-threat-food-security-humankind
11. Climate Change: Basic Information. Accessed through https://www.epa.gov/climatechange/climate-change-basic-information
12. Cao J: Beyond Copenhagen: Reconciling International Fairness, Economic Development, and Climate Protection. The Harvard Project on International Climate Agreements. 2010. Accessed through http://belfercenter.ksg.harvard.edu/files/CaoHPICADP44.pdf
13. Carius, Alexander, Dennis Tänzler and Judith Winterstein: Weltkarte von Umweltkonflikten – Ansätzeeiner Typologisierung. Externe Expertise für das WBGU: Hauptgutachten Welt im Wandel: Sicherheitsrisiko Klimawandel, 2006. Accessed through http://www.wbgu.de/wbgu_jg2007_ex02.pdf
14. Comparative Analysis of developing country national climate change strategies: Accessed through http://www.usclimatenetwork.org/

resource-database/WRI%20Comparative%20Analysis%20OF%20 Developing%20Country%2 0National%20Climate%20Change%20 Strategies.pdf

15. Could Trump simply withdraw US from Paris Climate Agreement: Accessed through https://www.scientificamerican.com/article/could-trump-simply-withdraw-u-s-from-paris-climate-agreement/
16. Department of Ecology State of Washington: What is Climate Change? Accessed through
17. Detraz N and Betsill M M: Climate Change and Environmental Security: For Whom the Discourse Shifts. International Studies Perspectives (2009) 10, 303–320, 2009 http://graduateinstitute.ch/files/live/sites/iheid/files/sites/political_science/shared/polit ical_science/9957/Detraz&Betsill%202009.pdf
18. Durban: at last, US, India and China agree to act on climate. Accessed through http://www.theweek.co.uk/environment/climate-change/43518/durban-last-us-india-and-china-agree-act-climate Dubash N.K.: Copenhagen: Climate of mistrust, Economic & Political Weekly 64 8–11, 2009
19. http://re.indiaenvironmentportal.org.in/files/COP.pdf.
20. http://www.ecy.wa.gov/climatechange/whatis.htm
21. http://www.epa.gov/climatechange/basics/
22. Factbox, What was agreed and left unfinished in UN climate deal. Accessed through http://in.reuters.com/article/idINIndia-44872920091220?sp=true
23. Fact Sheet: Overview of Clean Power. Accessed through https://www.epa.gov/cleanpowerplan/fact-sheet-overview-clean-power-plan
24. Giddens, Anthony, "The politics of climate change National responses to the challenge of global warming", pg.no.9 accessed through http://www.policy-network.net/uploadedFiles/Publications/Publications/The_politics_of_climate_change_Anthony_Giddens.pdf
25. Global Greenhouse Gas emissions data: Accessed through https://www.epa.gov/ghgemissions/global-greenhouse-gas-emissions-data
26. Learn about carbon pollution from power plants. Accessed through https://www.epa.gov/cleanpowerplan/learn-about-carbon-pollution-power-plants#what
27. http://www.ipcc.ch/publications_and_data/ar4/wg3/en/tssts-ts-13-4-international-agreements.html
28. Haftendorn, Helga: The Security Puzzle: Theory-Building and Discipline-Building in International Security. International Studies Quarterly (Blackwell Publishing on behalf of The International Studies Association) 35 (1): 3–17. doi:10.2307/2600386. JSTOR 2600386, 1991.

Accessed through http://en.wikipedia.org/wiki/Deterrence_%28 legal%29

29. IPCC: The intergovernmental panel on climate change, Working Group III Climate Change 2007: Mitigation of climate change May 2007. Accessed through http://www.ipcc.ch/pdf/assessment-report/ar4/wg3/ar4_wg3_full_report.pdf
30. IEA International Energy Agency CO_2 emissions from fuel combustion – highlights, 2013. Accessed through http://www.iea.org/co2highlights/co2highlights.pdf
31. India–United States Relations, Accessed through https://www.mea.gov.in/Portal/ForeignRelation/USA_15_01_2016.pdf
32. The UN Environmental Programme in climate Change and International Security Source: Population and Development Review, vol.37, No.3 September 2011 Published by: Population Council. Accessed through URL://www.jstor.org/stable/23036068 Page No. 601-606.
33. The UNFCCCs Durban Platform. Accessed through http://www.hfw.com/downloads/Client%20Brief%20-%20UNFCCCs%20Durban%20%5BA4%204pp%5D%20January%202012.pdf
34. National Action Plan on Climate Change: Accessed through http://www.moef.nic.in/modules/about-the-ministry/CCD/NAP_E.pdf page 2,
35. NOAA Climate Services. Accessed through https://www.ncdc.noaa.gov/oa/about/response-v2.pdf
36. Climate Change and the Kyoto Protocol—The Kyoto Protocol, December 11, 1997, Accessed through http://climatechange.sea.ca/kyoto_protocol.html
37. climatechange/background/general/rough_guide_unfccc_and_kyoto_protocol_090520.pdf http://nrg4sd.org/sites/default/files/default/files/content/public/29-
38. Lt. General RK Jasbir Singh, Indian Defence Yearbook, Natraj Publishers, New Delhi, 1997, pp. 13-25., 1997-1998. Accessed through http://fas.org/irp/nic/battilega/india.pdf\
39. Jones Bruce and Saram Samir, "An 'India Exception' and India-U.S. Partnership on Climate Change, Planet Policy," Monday, January 12, 2015. Accessed through https://www.brookings.edu/blog/planetpolicy/2015/01/12/an-india-exception-and-u-s-partnership-on-climate-change/
40. Kyoto Protocol Fast Facts, CNN Library Updated 1712 GMT (0112 HKT) March 30, 2016. Accessed through http://edition.cnn.com/2013/07/26/world/kyoto-protocol-fast-facts/
41. Outcomes of the U.N. Climate Change Conference in Paris 21st Session

of the Conference of the Parties to the United Nations Framework Convention on Climate Change (CO 21) November 30-December 12, 2015. Accessed through https://www.c2es.org/international/negotiations/cop21-paris/summary

42. Risbey J: The new climate discourse: Alarmist or alarming? Global Environmental Change 18, 2008 26–37. Accessed through http://www.marine.csiro.au/~ris009/pubfiles/gec_alarming.pdf
43. Subrahmanyam K., "Perspective in defence planning", Abhinav publications, New Delhi, 1972. Thaker J and LeiserowitzA (2014) Shifting discourses of climate change in India, Climatic Change DOI 10.1007/s10584-014-1059-6. Accessed through file:///E:/FINAL%20 REPORT/Chapter%202/Thaker_and_Leiserowitz_(2014)_Shifting_discourses_of_climate_change_in_India.pdf
44. Summary of the Clean Air Act: Accessed through https://www.epa.gov/laws-regulations/summary-clean-air-act
45. Senate confirms Scott Pruitt as EPA head. Accessed through https://www.nytimes.com/2017/02/17/us/politics/scott-pruitt-environmental-protection-agency.html?_r=0
46. Pellissery Sony, "India and the Politics of Climate Change", Institute of Rural Management, Anand Gujarat, India. Accessed through http://www.iddri.org/Evenements/Conferences/100323_presentation_IndiaandthPolitics% 20of%20Climate%20Change.pdf
47. https://www.quaker.org.uk/files/Climatechange-5-cooperation.pdf
48. The Paris Agreement: Accessed through http://unfccc.int/paris_agreement/items/9485.php
49. The Kyoto Protocol: History. Accessed through https://www.mtholyoke.edu/~danov20d/site/history.htm
50. The Paris Agreement explained. Accessed through http://www.abc.net.au/news/2016-12-09/the-paris-agreement-explained/8107100
51. The state of the climate in 2016, www.economist.com/blogs/graphicdetail/2016/11/global-warming accessed
52. The Kyoto Protocol: Environmental Protection Agency, Ireland. Accessed through http://www.epa.ie/climate/thekyotoprotocol/
53. Tackling the Global Climate Challenge. Accessed through http://www.unfoundation.org/what-we-do/issues/energy-and-climate/tackling-global-climate-challenge.html
54. The UN Environmental Programme in climate Change and International Security Source: Population and Development Review, vol.37, No.3 September 2011 Published by: Population Council. Accessed through URL://www.jstor.org/stable/23036068 Page No. 601-606
55. The Copenhagen Accord: Foundations for International Action on

Climate Change. Accessed through https://www.nrdc.org/sites/default/files/Copenhagen%20accord.pdf

56. UN Climate talks end with late deal. Accessed through http://www.bbc.com/news/science-environment-16124670
57. US State Department. Accessed through www.state.gov/r/pa/prs/ps/2015/09/247192.html
58. United Nations Framework Convention on Climate Change, Summary of the Paris Agreement. Accessed through http://bigpicture.unfccc.int/#content-the-paris-agreement
59. United Nations Framework Convention On Climate Change,5th October 2016. Accessed through http://unfccc.int/kyoto_protocol/items/2830.php
60. United Nations Framework Convention on Climate Change, FCCC/INFORMAL/84 GE.05-62220 (E) 200705, 1992. Accessed through http://unfccc.int/resource/docs/convkp/conveng.pdf
61. United Nations Framework Convention on Climate Change. Accessed Through https://en.wikipedia.org/wiki/United_Nations_Framework_ Convention_on_Climate_ Change
62. United Nations Framework Convention On Climate Change, 5thOctober2016. Accessed through http://unfccc.int/essential_background/convention/items/6036.php
63. United Nations Framework Convention on Climate Change. Accessed through http://unfccc.int/key_steps/bali_road_map/items/6072.php
64. UNDP and Climate Change Zero Carbon, Sustainable Development. Accessed through http://www.undp.org/content/dam/undp/library/Climate%20and%20Disaster%20Resi lience/UNDP_and_Climate_Change.pdf http://unfccc.int/kyoto_protocol/items/2830.php
65. United Nations Framework Convention on Climate Change. Accessed through http://unfccc.int/kyoto_protocol/background/items/2879.php Page No. 1.
66. United Nations Framework Convention on Climate Change. Accessed through http://unfccc.int/adaptation/items/5852.phn
67. United States Environmental Protection Agency Our Mission and what we do. Accessed through https://www.epa.gov/aboutepa/our-mission-and-what-we-do
68. UNEP 2007Annual Report. Accessed through http://www.unep.org/PDF/AnnualReport/2007/AnnualReport2007_en_web.pdf
69. UNFCCC (2006) United Nations Framework Convention on Climate Change: Handbook. Bonn, Germany: Climate Change Secretariat. Accessed through http://unfccc.int/resource/docs/publications/handbook.pdf
70. UNFCCC Report on Climate Change: Impacts, Vulnerabilities and

Adaptation in Developing Countries, 2007. Accessed through http://unfccc.int/resource/docs/publications/impacts.pdf

71. United States Environmental Protection Agency: Climate Change Science Facts, Office of Air and Radiation (6207J) EPA 430-F-10-002, 2010. Accessed through, http://www.epa.gov/climatechange/Downloads/Climate_Change_Science_Facts.pdf
72. Varad Pande : Changing the Climate of the India-U.S. Strategic Partnership 9th October, 2013, Accessed through http://www.business-standard.com/article/opinion/varad-pande-changing-the-climate-of-the-india-us-strategic-partnership-113101200638_1.html
73. Victor D.G.: Toward Effective International Cooperation on Climate Change: Numbers, Interests and Institutions, Global Environmental Politics 6:3, August 2006 by the Massachusetts Institute of Technology, 2006. Accessed through http://www.uio.no/studier/emner/annet/sum/SUM4015/h08/Victor.pdf on
74. Varghese K. George, India a partner in Climate Change fight: US. Accessed through www.thehindu.com/news/international/india-partner-in-climate-change-talks-says-us/article7930177ece 19 on April 2018
75. White house press office. Accessed through www.whitehouse.gov/the-press-office/2016/06/07 on 5 December 2019.
76. What is the Kyoto Protocol? -Carbonify. Accessed through www.carbonify.com/articles/kyoto-protocol.htm on 27 December 2016.
77. What Will Climate Change Mean for Human Security? Accessed through https://ourworld.unu.edu/en/what-will-climate-change-mean-for-human-security on 15 May 2019.
78. Wisner, Ben: Climate Change and Human Security. Accessed through, http://www.disasterdiplomacy.org/cchswisneretal.pdfon 15th April 2007.
79. What Cancun means for China and the US. Accessed through http://www.wri.org/blog/2010/12/what-cancun-means-china-and-us on 19 July2020.
80. What Scott Pruitt's confirmation means for the EPA. Accessed through http://www.vox.com/science-and-health/2017/2/17/14630614/scott-pruitt-epa-climate on 20 November 2018.

Index

*